Simplify each expression.

$2d^5 \cdot 4d^3 =$

$4g^5 \cdot 5g^3 =$

$5b \cdot 10b^3 =$

$5x^3 \cdot 10x^2 =$

$4b^3 \cdot 4b^2 =$

$10b^4 \cdot 3b^6 =$

$3v^6 \cdot 2v^2 =$

$5d^3 \cdot 3d^6 =$

$4d^5 \cdot 2d^5 =$

$2m \cdot 4m =$

$2t^3 \cdot 3t^3 =$

$2a^5 \cdot 3a^4 =$

$5m^2 \cdot 4m^2 =$

$10f^5 \cdot 2f^4 =$

$5t^4 \cdot 5t^2 =$

$5y^5 \cdot 4y^6 =$

$4g^3 \cdot 3g^2 =$

$3y^2 \cdot 10y^5 =$

$4f^4 \cdot 3f^2 =$

$3a^6 \cdot 4a^6 =$

# Answer Key

Simplify each expression.

$2d^5 \cdot 4d^3 = 8d^8$

$4g^5 \cdot 5g^3 = 20g^8$

$5b \cdot 10b^3 = 50b^4$

$5x^3 \cdot 10x^2 = 50x^5$

$4b^3 \cdot 4b^2 = 16b^5$

$10b^4 \cdot 3b^6 = 30b^{10}$

$3v^6 \cdot 2v^2 = 6v^8$

$5d^3 \cdot 3d^6 = 15d^9$

$4d^5 \cdot 2d^5 = 8d^{10}$

$2m \cdot 4m = 8m^2$

$2t^3 \cdot 3t^3 = 6t^6$

$2a^5 \cdot 3a^4 = 6a^9$

$5m^2 \cdot 4m^2 = 20m^4$

$10f^5 \cdot 2f^4 = 20f^9$

$5t^4 \cdot 5t^2 = 25t^6$

$5y^5 \cdot 4y^6 = 20y^{11}$

$4g^3 \cdot 3g^2 = 12g^5$

$3y^2 \cdot 10y^5 = 30y^7$

$4f^4 \cdot 3f^2 = 12f^6$

$3a^6 \cdot 4a^6 = 12a^{12}$

Simplify each expression.

$2w^4 \cdot 3w^5 =$

$4g^5 \cdot 3g =$

$10f^4 \cdot 4f^3 =$

$5b \cdot 4b =$

$2g^4 \cdot 10g^4 =$

$3g \cdot 10g^5 =$

$2k^5 \cdot 10k^2 =$

$4k^6 \cdot 3k =$

$3k \cdot 3k^6 =$

$5a^2 \cdot 5a^6 =$

$2k^6 \cdot 4k =$

$5x^5 \cdot 10x^2 =$

$5b^4 \cdot 3b^3 =$

$3n^4 \cdot 5n^3 =$

$3y^2 \cdot 3y^3 =$

$5h^6 \cdot 3h^5 =$

$5h^2 \cdot 5h^6 =$

$10u^2 \cdot 2u^5 =$

$3t \cdot 5t^6 =$

$5m^2 \cdot 4m =$

# Answer Key

Simplify each expression.

$2w^4 \cdot 3w^5 = 6w^9$

$4g^5 \cdot 3g = 12g^6$

$10f^4 \cdot 4f^3 = 40f^7$

$5b \cdot 4b = 20b^2$

$2g^4 \cdot 10g^4 = 20g^8$

$3g \cdot 10g^5 = 30g^6$

$2k^5 \cdot 10k^2 = 20k^7$

$4k^6 \cdot 3k = 12k^7$

$3k \cdot 3k^6 = 9k^7$

$5a^2 \cdot 5a^6 = 25a^8$

$2k^6 \cdot 4k = 8k^7$

$5x^5 \cdot 10x^2 = 50x^7$

$5b^4 \cdot 3b^3 = 15b^7$

$3n^4 \cdot 5n^3 = 15n^7$

$3y^2 \cdot 3y^3 = 9y^5$

$5h^6 \cdot 3h^5 = 15h^{11}$

$5h^2 \cdot 5h^6 = 25h^8$

$10u^2 \cdot 2u^5 = 20u^7$

$3t \cdot 5t^6 = 15t^7$

$5m^2 \cdot 4m = 20m^3$

Simplify each expression.

$4d^4 \cdot 4d^2 =$

$2m^4 \cdot 10m^4 =$

$5p^6 \cdot 10p^6 =$

$10a^4 \cdot 3a^4 =$

$2q^2 \cdot 2q =$

$5p^2 \cdot 5p =$

$2q \cdot 3q^3 =$

$10t \cdot 5t^4 =$

$3x^5 \cdot 3x =$

$10a^2 \cdot 2a^6 =$

$2b^5 \cdot 5b^6 =$

$4b^5 \cdot 2b^2 =$

$4b^4 \cdot 3b^4 =$

$4f^2 \cdot 2f^5 =$

$2y^6 \cdot 3y^6 =$

$4x \cdot 3x^3 =$

$3k^6 \cdot 2k^5 =$

$10m^3 \cdot 5m =$

$10y^2 \cdot 5y^3 =$

$2d^5 \cdot 4d^5 =$

# Answer Key

Simplify each expression.

$4d^4 \cdot 4d^2 = 16d^6$

$2m^4 \cdot 10m^4 = 20m^8$

$5p^6 \cdot 10p^6 = 50p^{12}$

$10a^4 \cdot 3a^4 = 30a^8$

$2q^2 \cdot 2q = 4q^3$

$5p^2 \cdot 5p = 25p^3$

$2q \cdot 3q^3 = 6q^4$

$10t \cdot 5t^4 = 50t^5$

$3x^5 \cdot 3x = 9x^6$

$10a^2 \cdot 2a^6 = 20a^8$

$2b^5 \cdot 5b^6 = 10b^{11}$

$4b^5 \cdot 2b^2 = 8b^7$

$4b^4 \cdot 3b^4 = 12b^8$

$4f^2 \cdot 2f^5 = 8f^7$

$2y^6 \cdot 3y^6 = 6y^{12}$

$4x \cdot 3x^3 = 12x^4$

$3k^6 \cdot 2k^5 = 6k^{11}$

$10m^3 \cdot 5m = 50m^4$

$10y^2 \cdot 5y^3 = 50y^5$

$2d^5 \cdot 4d^5 = 8d^{10}$

Simplify each expression.

$3a^2 \cdot 4a^2 =$

$5v \cdot 10v =$

$2a^2 \cdot 4a^6 =$

$2w^3 \cdot 3w^3 =$

$5f^5 \cdot 4f^6 =$

$4w^3 \cdot 4w =$

$3y^5 \cdot 4y^6 =$

$5b^3 \cdot 10b =$

$2b^6 \cdot 2b^5 =$

$5k^6 \cdot 3k^5 =$

$5a^3 \cdot 3a^4 =$

$5y^5 \cdot 2y^6 =$

$2k^2 \cdot 2k^2 =$

$2a \cdot 4a^6 =$

$10y^3 \cdot 10y^3 =$

$3p^6 \cdot 3p^3 =$

$10v^3 \cdot 10v^4 =$

$4t^2 \cdot 4t^6 =$

$3f^5 \cdot 2f^6 =$

$5f^2 \cdot 2f^4 =$

# Answer Key

Simplify each expression.

$3a^2 \cdot 4a^2 = 12a^4$

$5v \cdot 10v = 50v^2$

$2a^2 \cdot 4a^6 = 8a^8$

$2w^3 \cdot 3w^3 = 6w^6$

$5f^5 \cdot 4f^6 = 20f^{11}$

$4w^3 \cdot 4w = 16w^4$

$3y^5 \cdot 4y^6 = 12y^{11}$

$5b^3 \cdot 10b = 50b^4$

$2b^6 \cdot 2b^5 = 4b^{11}$

$5k^6 \cdot 3k^5 = 15k^{11}$

$5a^3 \cdot 3a^4 = 15a^7$

$5y^5 \cdot 2y^6 = 10y^{11}$

$2k^2 \cdot 2k^2 = 4k^4$

$2a \cdot 4a^6 = 8a^7$

$10y^3 \cdot 10y^3 = 100y^6$

$3p^6 \cdot 3p^3 = 9p^9$

$10v^3 \cdot 10v^4 = 100v^7$

$4t^2 \cdot 4t^6 = 16t^8$

$3f^5 \cdot 2f^6 = 6f^{11}$

$5f^2 \cdot 2f^4 = 10f^6$

Simplify each expression.

$2w^3 \cdot 10w^3 =$

$10g \cdot 2g^3 =$

$5p^6 \cdot 2p^2 =$

$3b^6 \cdot 5b^3 =$

$4d^6 \cdot 4d =$

$2b^5 \cdot 5b =$

$2y^5 \cdot 2y^2 =$

$5r^4 \cdot 3r^6 =$

$2w^4 \cdot 2w^2 =$

$4p^3 \cdot 2p^3 =$

$5h^5 \cdot 2h =$

$2d^6 \cdot 5d^5 =$

$3n^4 \cdot 2n^4 =$

$2r^6 \cdot 5r^2 =$

$2u^2 \cdot 10u =$

$5h^6 \cdot 10h^6 =$

$3f^5 \cdot 4f^2 =$

$4t^5 \cdot 5t =$

$4u \cdot 4u^4 =$

$3t \cdot 4t =$

# Answer Key

Simplify each expression.

$2w^3 \cdot 10w^3 = 20w^6$

$10g \cdot 2g^3 = 20g^4$

$5p^6 \cdot 2p^2 = 10p^8$

$3b^6 \cdot 5b^3 = 15b^9$

$4d^6 \cdot 4d = 16d^7$

$2b^5 \cdot 5b = 10b^6$

$2y^5 \cdot 2y^2 = 4y^7$

$5r^4 \cdot 3r^6 = 15r^{10}$

$2w^4 \cdot 2w^2 = 4w^6$

$4p^3 \cdot 2p^3 = 8p^6$

$5h^5 \cdot 2h = 10h^6$

$2d^6 \cdot 5d^5 = 10d^{11}$

$3n^4 \cdot 2n^4 = 6n^8$

$2r^6 \cdot 5r^2 = 10r^8$

$2u^2 \cdot 10u = 20u^3$

$5h^6 \cdot 10h^6 = 50h^{12}$

$3f^5 \cdot 4f^2 = 12f^7$

$4t^5 \cdot 5t = 20t^6$

$4u \cdot 4u^4 = 16u^5$

$3t \cdot 4t = 12t^2$

Simplify each expression.

$3k \cdot 3k^4 =$

$2p^4 \cdot 2p^5 =$

$3k^6 \cdot 10k^2 =$

$2a^4 \cdot 5a^2 =$

$10p^2 \cdot 4p =$

$5m^3 \cdot 4m^5 =$

$10y^6 \cdot 10y^3 =$

$3x^5 \cdot 2x^6 =$

$3w^5 \cdot 3w^6 =$

$3h^6 \cdot 10h^4 =$

$2p \cdot 3p^5 =$

$2b^5 \cdot 10b =$

$3p^2 \cdot 5p^6 =$

$10y^4 \cdot 3y^5 =$

$10y^2 \cdot 4y^3 =$

$10m^2 \cdot 3m^2 =$

$4k^5 \cdot 2k^5 =$

$10y^3 \cdot 3y^5 =$

$10p^4 \cdot 3p^2 =$

$2k^3 \cdot 4k^3 =$

# Answer Key

Simplify each expression.

$3k \cdot 3k^4 = 9k^5$

$2p^4 \cdot 2p^5 = 4p^9$

$3k^6 \cdot 10k^2 = 30k^8$

$2a^4 \cdot 5a^2 = 10a^6$

$10p^2 \cdot 4p = 40p^3$

$5m^3 \cdot 4m^5 = 20m^8$

$10y^6 \cdot 10y^3 = 100y^9$

$3x^5 \cdot 2x^6 = 6x^{11}$

$3w^5 \cdot 3w^6 = 9w^{11}$

$3h^6 \cdot 10h^4 = 30h^{10}$

$2p \cdot 3p^5 = 6p^6$

$2b^5 \cdot 10b = 20b^6$

$3p^2 \cdot 5p^6 = 15p^8$

$10y^4 \cdot 3y^5 = 30y^9$

$10y^2 \cdot 4y^3 = 40y^5$

$10m^2 \cdot 3m^2 = 30m^4$

$4k^5 \cdot 2k^5 = 8k^{10}$

$10y^3 \cdot 3y^5 = 30y^8$

$10p^4 \cdot 3p^2 = 30p^6$

$2k^3 \cdot 4k^3 = 8k^6$

Simplify each expression.

$4p^6 \cdot 3p^3 =$

$3u^4 \cdot 2u^6 =$

$5x^2 \cdot 2x^4 =$

$10u^6 \cdot 4u^4 =$

$4h^4 \cdot 10h^2 =$

$5b^2 \cdot 2b^2 =$

$3u \cdot 4u^6 =$

$3u^6 \cdot 5u^5 =$

$5w^2 \cdot 4w^4 =$

$2x^5 \cdot 4x^3 =$

$10w^6 \cdot 5w =$

$10h^4 \cdot 4h^4 =$

$10k^5 \cdot 3k^2 =$

$4k^6 \cdot 3k^3 =$

$3m^5 \cdot 4m^2 =$

$4n^3 \cdot 5n^4 =$

$4h^5 \cdot 5h^5 =$

$5t^6 \cdot 4t^3 =$

$2n^2 \cdot 4n^3 =$

$4f^5 \cdot 2f^5 =$

# Answer Key

Simplify each expression.

$4p^6 \cdot 3p^3 = 12p^9$

$3u^4 \cdot 2u^6 = 6u^{10}$

$5x^2 \cdot 2x^4 = 10x^6$

$10u^6 \cdot 4u^4 = 40u^{10}$

$4h^4 \cdot 10h^2 = 40h^6$

$5b^2 \cdot 2b^2 = 10b^4$

$3u \cdot 4u^6 = 12u^7$

$3u^6 \cdot 5u^5 = 15u^{11}$

$5w^2 \cdot 4w^4 = 20w^6$

$2x^5 \cdot 4x^3 = 8x^8$

$10w^6 \cdot 5w = 50w^7$

$10h^4 \cdot 4h^4 = 40h^8$

$10k^5 \cdot 3k^2 = 30k^7$

$4k^6 \cdot 3k^3 = 12k^9$

$3m^5 \cdot 4m^2 = 12m^7$

$4n^3 \cdot 5n^4 = 20n^7$

$4h^5 \cdot 5h^5 = 20h^{10}$

$5t^6 \cdot 4t^3 = 20t^9$

$2n^2 \cdot 4n^3 = 8n^5$

$4f^5 \cdot 2f^5 = 8f^{10}$

Simplify each expression.

$4g \cdot 10g^3 =$

$5m^4 \cdot 2m^3 =$

$5a^4 \cdot 4a^6 =$

$3f^5 \cdot 2f^6 =$

$3k^6 \cdot 5k^2 =$

$10b^3 \cdot 3b^4 =$

$3v \cdot 3v^5 =$

$10g^3 \cdot 3g^6 =$

$10d \cdot 10d^5 =$

$4w^3 \cdot 5w^5 =$

$10x^2 \cdot 3x^3 =$

$2h^3 \cdot 5h^4 =$

$4w^4 \cdot 4w^6 =$

$3a^6 \cdot 2a^5 =$

$4k^5 \cdot 3k^6 =$

$4u^3 \cdot 3u^6 =$

$3g^2 \cdot 3g^3 =$

$4n^4 \cdot 3n^3 =$

$2h^3 \cdot 3h^5 =$

$3q^4 \cdot 3q =$

# Answer Key

Simplify each expression.

$4g \cdot 10g^3 = 40g^4$

$5m^4 \cdot 2m^3 = 10m^7$

$5a^4 \cdot 4a^6 = 20a^{10}$

$3f^5 \cdot 2f^6 = 6f^{11}$

$3k^6 \cdot 5k^2 = 15k^8$

$10b^3 \cdot 3b^4 = 30b^7$

$3v \cdot 3v^5 = 9v^6$

$10g^3 \cdot 3g^6 = 30g^9$

$10d \cdot 10d^5 = 100d^6$

$4w^3 \cdot 5w^5 = 20w^8$

$10x^2 \cdot 3x^3 = 30x^5$

$2h^3 \cdot 5h^4 = 10h^7$

$4w^4 \cdot 4w^6 = 16w^{10}$

$3a^6 \cdot 2a^5 = 6a^{11}$

$4k^5 \cdot 3k^6 = 12k^{11}$

$4u^3 \cdot 3u^6 = 12u^9$

$3g^2 \cdot 3g^3 = 9g^5$

$4n^4 \cdot 3n^3 = 12n^7$

$2h^3 \cdot 3h^5 = 6h^8$

$3q^4 \cdot 3q = 9q^5$

Simplify each expression.

$4r^4 \cdot 5r^4 =$

$5f^2 \cdot 2f =$

$5g^5 \cdot 3g^2 =$

$10y^2 \cdot 5y^2 =$

$10m^4 \cdot 2m^2 =$

$2w^4 \cdot 10w =$

$5x^3 \cdot 5x^2 =$

$4p^2 \cdot 2p^2 =$

$3k^3 \cdot 4k^3 =$

$5y^4 \cdot 10y^3 =$

$10h^2 \cdot 4h =$

$2t^5 \cdot 2t^5 =$

$3y^2 \cdot 10y^2 =$

$4n^2 \cdot 2n^3 =$

$3p^5 \cdot 2p^4 =$

$4p^6 \cdot 4p^4 =$

$4k^3 \cdot 2k =$

$2p^3 \cdot 5p^5 =$

$5f^3 \cdot 3f =$

$2g^2 \cdot 3g =$

# Answer Key

Simplify each expression.

$4r^4 \cdot 5r^4 = 20r^8$

$5f^2 \cdot 2f = 10f^3$

$5g^5 \cdot 3g^2 = 15g^7$

$10y^2 \cdot 5y^2 = 50y^4$

$10m^4 \cdot 2m^2 = 20m^6$

$2w^4 \cdot 10w = 20w^5$

$5x^3 \cdot 5x^2 = 25x^5$

$4p^2 \cdot 2p^2 = 8p^4$

$3k^3 \cdot 4k^3 = 12k^6$

$5y^4 \cdot 10y^3 = 50y^7$

$10h^2 \cdot 4h = 40h^3$

$2t^5 \cdot 2t^5 = 4t^{10}$

$3y^2 \cdot 10y^2 = 30y^4$

$4n^2 \cdot 2n^3 = 8n^5$

$3p^5 \cdot 2p^4 = 6p^9$

$4p^6 \cdot 4p^4 = 16p^{10}$

$4k^3 \cdot 2k = 8k^4$

$2p^3 \cdot 5p^5 = 10p^8$

$5f^3 \cdot 3f = 15f^4$

$2g^2 \cdot 3g = 6g^3$

Simplify each expression.

$10n^4 \cdot 3n^3 =$

$3v^3 \cdot 4v^2 =$

$10y^4 \cdot 10y^4 =$

$4m^3 \cdot 10m^3 =$

$5d \cdot 10d^5 =$

$4y^6 \cdot 3y^3 =$

$5q^4 \cdot 5q^3 =$

$4r^2 \cdot 10r^2 =$

$5r \cdot 4r =$

$4r^2 \cdot 10r^6 =$

$5t^2 \cdot 3t^4 =$

$5a^5 \cdot 5a =$

$4b^2 \cdot 10b^2 =$

$2k^2 \cdot 5k^4 =$

$5u^5 \cdot 3u^5 =$

$2u^5 \cdot 5u^3 =$

$4v \cdot 3v^5 =$

$4y^4 \cdot 4y^5 =$

$3k^5 \cdot 2k^6 =$

$3u^2 \cdot 10u^4 =$

# Answer Key

Simplify each expression.

$10n^4 \cdot 3n^3 = 30n^7$

$10y^4 \cdot 10y^4 = 100y^8$

$5d \cdot 10d^5 = 50d^6$

$5q^4 \cdot 5q^3 = 25q^7$

$5r \cdot 4r = 20r^2$

$5t^2 \cdot 3t^4 = 15t^6$

$4b^2 \cdot 10b^2 = 40b^4$

$5u^5 \cdot 3u^5 = 15u^{10}$

$4v \cdot 3v^5 = 12v^6$

$3k^5 \cdot 2k^6 = 6k^{11}$

$3v^3 \cdot 4v^2 = 12v^5$

$4m^3 \cdot 10m^3 = 40m^6$

$4y^6 \cdot 3y^3 = 12y^9$

$4r^2 \cdot 10r^2 = 40r^4$

$4r^2 \cdot 10r^6 = 40r^8$

$5a^5 \cdot 5a = 25a^6$

$2k^2 \cdot 5k^4 = 10k^6$

$2u^5 \cdot 5u^3 = 10u^8$

$4y^4 \cdot 4y^5 = 16y^9$

$3u^2 \cdot 10u^4 = 30u^6$

Simplify each expression.

$3g \cdot 2g^4 =$

$5a^3 \cdot 10a^5 =$

$3r \cdot 3r =$

$4w^3 \cdot 3w^6 =$

$5q^4 \cdot 5q^6 =$

$2r^3 \cdot 5r^6 =$

$3p^6 \cdot 4p^2 =$

$3f^3 \cdot 2f^5 =$

$4w^4 \cdot 5w^2 =$

$5h^4 \cdot 4h^5 =$

$5t^5 \cdot 3t^3 =$

$4t^2 \cdot 4t^4 =$

$3t^3 \cdot 5t^4 =$

$3m^5 \cdot 3m^3 =$

$2w^3 \cdot 2w^6 =$

$5r \cdot 2r^5 =$

$5a^5 \cdot 5a^4 =$

$2y^3 \cdot 4y^3 =$

$3a \cdot 2a^6 =$

$3r^4 \cdot 4r^6 =$

# Answer Key

Simplify each expression.

$3g \cdot 2g^4 = 6g^5$

$5a^3 \cdot 10a^5 = 50a^8$

$3r \cdot 3r = 9r^2$

$4w^3 \cdot 3w^6 = 12w^9$

$5q^4 \cdot 5q^6 = 25q^{10}$

$2r^3 \cdot 5r^6 = 10r^9$

$3p^6 \cdot 4p^2 = 12p^8$

$3f^3 \cdot 2f^5 = 6f^8$

$4w^4 \cdot 5w^2 = 20w^6$

$5h^4 \cdot 4h^5 = 20h^9$

$5t^5 \cdot 3t^3 = 15t^8$

$4t^2 \cdot 4t^4 = 16t^6$

$3t^3 \cdot 5t^4 = 15t^7$

$3m^5 \cdot 3m^3 = 9m^8$

$2w^3 \cdot 2w^6 = 4w^9$

$5r \cdot 2r^5 = 10r^6$

$5a^5 \cdot 5a^4 = 25a^9$

$2y^3 \cdot 4y^3 = 8y^6$

$3a \cdot 2a^6 = 6a^7$

$3r^4 \cdot 4r^6 = 12r^{10}$

Simplify each expression.

$5p^5 \cdot 3p =$

$5u^2 \cdot 10u^4 =$

$3d^4 \cdot 3d^3 =$

$2k^3 \cdot 2k^2 =$

$5r^4 \cdot 5r^3 =$

$10g^4 \cdot 3g^2 =$

$5q^6 \cdot 4q^6 =$

$3b^2 \cdot 2b^4 =$

$5d^4 \cdot 5d^5 =$

$5q \cdot 3q =$

$10b^3 \cdot 3b^2 =$

$10t^3 \cdot 3t^2 =$

$5n^4 \cdot 5n =$

$4f^2 \cdot 5f^5 =$

$4g^2 \cdot 4g^3 =$

$3q^4 \cdot 5q^4 =$

$5n^6 \cdot 2n^5 =$

$3w^4 \cdot 2w^4 =$

$10r^4 \cdot 5r^5 =$

$4n^2 \cdot 4n^6 =$

# Answer Key

Simplify each expression.

$5p^5 \cdot 3p = 15p^6$

$10b^3 \cdot 3b^2 = 30b^5$

$5u^2 \cdot 10u^4 = 50u^6$

$10t^3 \cdot 3t^2 = 30t^5$

$3d^4 \cdot 3d^3 = 9d^7$

$5n^4 \cdot 5n = 25n^5$

$2k^3 \cdot 2k^2 = 4k^5$

$4f^2 \cdot 5f^5 = 20f^7$

$5r^4 \cdot 5r^3 = 25r^7$

$4g^2 \cdot 4g^3 = 16g^5$

$10g^4 \cdot 3g^2 = 30g^6$

$3q^4 \cdot 5q^4 = 15q^8$

$5q^6 \cdot 4q^6 = 20q^{12}$

$5n^6 \cdot 2n^5 = 10n^{11}$

$3b^2 \cdot 2b^4 = 6b^6$

$3w^4 \cdot 2w^4 = 6w^8$

$5d^4 \cdot 5d^5 = 25d^9$

$10r^4 \cdot 5r^5 = 50r^9$

$5q \cdot 3q = 15q^2$

$4n^2 \cdot 4n^6 = 16n^8$

Simplify each expression.

$4n \cdot 4n =$

$10m^3 \cdot 5m^3 =$

$4v \cdot 3v^3 =$

$2p^5 \cdot 5p =$

$4w \cdot 10w^4 =$

$3g^2 \cdot 5g^5 =$

$5w^2 \cdot 2w =$

$5r \cdot 3r^5 =$

$4k^5 \cdot 4k^4 =$

$3b^4 \cdot 2b^2 =$

$5w^5 \cdot 2w^3 =$

$4p^5 \cdot 10p =$

$10m^2 \cdot 5m =$

$3v^2 \cdot 5v^6 =$

$4p^2 \cdot 5p^5 =$

$4k^3 \cdot 3k^4 =$

$2p^2 \cdot 3p^2 =$

$2t^4 \cdot 3t^2 =$

$2u^5 \cdot 2u =$

$10h^5 \cdot 4h^3 =$

# Answer Key

Simplify each expression.

$4n \cdot 4n = 16n^2$

$10m^3 \cdot 5m^3 = 50m^6$

$4v \cdot 3v^3 = 12v^4$

$2p^5 \cdot 5p = 10p^6$

$4w \cdot 10w^4 = 40w^5$

$3g^2 \cdot 5g^5 = 15g^7$

$5w^2 \cdot 2w = 10w^3$

$5r \cdot 3r^5 = 15r^6$

$4k^5 \cdot 4k^4 = 16k^9$

$3b^4 \cdot 2b^2 = 6b^6$

$5w^5 \cdot 2w^3 = 10w^8$

$4p^5 \cdot 10p = 40p^6$

$10m^2 \cdot 5m = 50m^3$

$3v^2 \cdot 5v^6 = 15v^8$

$4p^2 \cdot 5p^5 = 20p^7$

$4k^3 \cdot 3k^4 = 12k^7$

$2p^2 \cdot 3p^2 = 6p^4$

$2t^4 \cdot 3t^2 = 6t^6$

$2u^5 \cdot 2u = 4u^6$

$10h^5 \cdot 4h^3 = 40h^8$

Simplify each expression.

$10y^6 \cdot 3y^2 =$

$2m \cdot 5m^3 =$

$3y^6 \cdot 5y =$

$10m^3 \cdot 5m^5 =$

$2r^3 \cdot 10r =$

$3u^5 \cdot 10u^3 =$

$2x^2 \cdot 5x^6 =$

$4h \cdot 4h^5 =$

$3m^5 \cdot 2m^3 =$

$3y^2 \cdot 4y^2 =$

$10f^5 \cdot 2f^5 =$

$2h^4 \cdot 2h^5 =$

$2v^4 \cdot 5v =$

$3n^2 \cdot 2n^3 =$

$3p^3 \cdot 3p^2 =$

$5y \cdot 10y^6 =$

$4g^6 \cdot 4g^3 =$

$3d \cdot 10d^4 =$

$10u^4 \cdot 4u^5 =$

$3b^6 \cdot 3b^3 =$

# Answer Key

Simplify each expression.

$10y^6 \cdot 3y^2 = 30y^8$

$2m \cdot 5m^3 = 10m^4$

$3y^6 \cdot 5y = 15y^7$

$10m^3 \cdot 5m^5 = 50m^8$

$2r^3 \cdot 10r = 20r^4$

$3u^5 \cdot 10u^3 = 30u^8$

$2x^2 \cdot 5x^6 = 10x^8$

$4h \cdot 4h^5 = 16h^6$

$3m^5 \cdot 2m^3 = 6m^8$

$3y^2 \cdot 4y^2 = 12y^4$

$10f^5 \cdot 2f^5 = 20f^{10}$

$2h^4 \cdot 2h^5 = 4h^9$

$2v^4 \cdot 5v = 10v^5$

$3n^2 \cdot 2n^3 = 6n^5$

$3p^3 \cdot 3p^2 = 9p^5$

$5y \cdot 10y^6 = 50y^7$

$4g^6 \cdot 4g^3 = 16g^9$

$3d \cdot 10d^4 = 30d^5$

$10u^4 \cdot 4u^5 = 40u^9$

$3b^6 \cdot 3b^3 = 9b^9$

Simplify each expression.

$4x^5 \cdot 3x^5 =$

$2v^5 \cdot 5v =$

$3k \cdot 3k^6 =$

$10p^6 \cdot 10p^6 =$

$3u^3 \cdot 10u^2 =$

$4r^5 \cdot 2r^2 =$

$3t^5 \cdot 10t^3 =$

$2r^5 \cdot 2r^3 =$

$10n^6 \cdot 4n^5 =$

$2r \cdot 5r^4 =$

$10r^3 \cdot 5r^4 =$

$10w^2 \cdot 3w^2 =$

$3m^5 \cdot 2m^3 =$

$3m^3 \cdot 10m^3 =$

$5g^3 \cdot 10g^3 =$

$2h^4 \cdot 10h^6 =$

$3f^3 \cdot 2f =$

$4x^2 \cdot 2x =$

$2x \cdot 3x^3 =$

$5t^5 \cdot 4t^4 =$

# Answer Key

Simplify each expression.

$4x^5 \cdot 3x^5 = 12x^{10}$

$2v^5 \cdot 5v = 10v^6$

$3k \cdot 3k^6 = 9k^7$

$10p^6 \cdot 10p^6 = 100p^{12}$

$3u^3 \cdot 10u^2 = 30u^5$

$4r^5 \cdot 2r^2 = 8r^7$

$3t^5 \cdot 10t^3 = 30t^8$

$2r^5 \cdot 2r^3 = 4r^8$

$10n^6 \cdot 4n^5 = 40n^{11}$

$2r \cdot 5r^4 = 10r^5$

$10r^3 \cdot 5r^4 = 50r^7$

$10w^2 \cdot 3w^2 = 30w^4$

$3m^5 \cdot 2m^3 = 6m^8$

$3m^3 \cdot 10m^3 = 30m^6$

$5g^3 \cdot 10g^3 = 50g^6$

$2h^4 \cdot 10h^6 = 20h^{10}$

$3f^3 \cdot 2f = 6f^4$

$4x^2 \cdot 2x = 8x^3$

$2x \cdot 3x^3 = 6x^4$

$5t^5 \cdot 4t^4 = 20t^9$

Simplify each expression.

$3n^4 \cdot 10n =$

$3w^5 \cdot 3w^2 =$

$4a \cdot 5a^3 =$

$5f^4 \cdot 4f^2 =$

$4u^5 \cdot 2u^5 =$

$4v^2 \cdot 3v^2 =$

$3r^4 \cdot 5r =$

$4w^2 \cdot 4w^3 =$

$4v^3 \cdot 2v =$

$4b^2 \cdot 3b^6 =$

$4m^6 \cdot 10m =$

$2b^3 \cdot 2b^2 =$

$2d^6 \cdot 2d =$

$5u^3 \cdot 2u^3 =$

$4q^6 \cdot 4q^3 =$

$3g^3 \cdot 4g^2 =$

$4t \cdot 5t^2 =$

$3y^4 \cdot 10y^4 =$

$5q^4 \cdot 2q^5 =$

$3h^5 \cdot 4h^2 =$

# Answer Key

Simplify each expression.

$3n^4 \cdot 10n = 30n^5$

$3w^5 \cdot 3w^2 = 9w^7$

$4a \cdot 5a^3 = 20a^4$

$5f^4 \cdot 4f^2 = 20f^6$

$4u^5 \cdot 2u^5 = 8u^{10}$

$4v^2 \cdot 3v^2 = 12v^4$

$3r^4 \cdot 5r = 15r^5$

$4w^2 \cdot 4w^3 = 16w^5$

$4v^3 \cdot 2v = 8v^4$

$4b^2 \cdot 3b^6 = 12b^8$

$4m^6 \cdot 10m = 40m^7$

$2b^3 \cdot 2b^2 = 4b^5$

$2d^6 \cdot 2d = 4d^7$

$5u^3 \cdot 2u^3 = 10u^6$

$4q^6 \cdot 4q^3 = 16q^9$

$3g^3 \cdot 4g^2 = 12g^5$

$4t \cdot 5t^2 = 20t^3$

$3y^4 \cdot 10y^4 = 30y^8$

$5q^4 \cdot 2q^5 = 10q^9$

$3h^5 \cdot 4h^2 = 12h^7$

Simplify each expression.

$2q^2 \cdot 10q^5 =$

$2v^5 \cdot 2v^2 =$

$5r \cdot 5r =$

$4v^2 \cdot 5v^3 =$

$2m^2 \cdot 5m^5 =$

$4v^3 \cdot 3v^2 =$

$4d^3 \cdot 3d^2 =$

$2d^3 \cdot 2d^6 =$

$10g^6 \cdot 3g^5 =$

$3w^4 \cdot 10w^3 =$

$10w \cdot 2w^2 =$

$5h^2 \cdot 2h^3 =$

$4m^2 \cdot 3m^3 =$

$5m^2 \cdot 3m^2 =$

$2y^2 \cdot 4y^2 =$

$3y^3 \cdot 10y =$

$10m^5 \cdot 3m^6 =$

$10n^2 \cdot 5n^2 =$

$3f^6 \cdot 3f^4 =$

$2k^3 \cdot 5k^6 =$

# Answer Key

Simplify each expression.

$2q^2 \cdot 10q^5 = 20q^7$

$2v^5 \cdot 2v^2 = 4v^7$

$5r \cdot 5r = 25r^2$

$4v^2 \cdot 5v^3 = 20v^5$

$2m^2 \cdot 5m^5 = 10m^7$

$4v^3 \cdot 3v^2 = 12v^5$

$4d^3 \cdot 3d^2 = 12d^5$

$2d^3 \cdot 2d^6 = 4d^9$

$10g^6 \cdot 3g^5 = 30g^{11}$

$3w^4 \cdot 10w^3 = 30w^7$

$10w \cdot 2w^2 = 20w^3$

$5h^2 \cdot 2h^3 = 10h^5$

$4m^2 \cdot 3m^3 = 12m^5$

$5m^2 \cdot 3m^2 = 15m^4$

$2y^2 \cdot 4y^2 = 8y^4$

$3y^3 \cdot 10y = 30y^4$

$10m^5 \cdot 3m^6 = 30m^{11}$

$10n^2 \cdot 5n^2 = 50n^4$

$3f^6 \cdot 3f^4 = 9f^{10}$

$2k^3 \cdot 5k^6 = 10k^9$

Simplify each expression.

$3d^4 \cdot 10d^4 =$

$4q^5 \cdot 10q^5 =$

$3f^4 \cdot 4f^2 =$

$3b^6 \cdot 3b^4 =$

$4n^2 \cdot 2n^3 =$

$4a^2 \cdot 2a^5 =$

$5h^5 \cdot 2h =$

$4g^2 \cdot 2g^6 =$

$5a^3 \cdot 3a^5 =$

$3t^2 \cdot 2t^6 =$

$5p^2 \cdot 3p^2 =$

$3k^2 \cdot 3k^3 =$

$4p \cdot 3p^4 =$

$4d^2 \cdot 3d^5 =$

$5t^2 \cdot 5t =$

$2q^4 \cdot 10q^3 =$

$10t^6 \cdot 5t^2 =$

$4n^2 \cdot 3n^2 =$

$10h^6 \cdot 10h^2 =$

$4v^5 \cdot 2v^6 =$

# Answer Key

Simplify each expression.

$3d^4 \cdot 10d^4 = 30d^8$

$4q^5 \cdot 10q^5 = 40q^{10}$

$3f^4 \cdot 4f^2 = 12f^6$

$3b^6 \cdot 3b^4 = 9b^{10}$

$4n^2 \cdot 2n^3 = 8n^5$

$4a^2 \cdot 2a^5 = 8a^7$

$5h^5 \cdot 2h = 10h^6$

$4g^2 \cdot 2g^6 = 8g^8$

$5a^3 \cdot 3a^5 = 15a^8$

$3t^2 \cdot 2t^6 = 6t^8$

$5p^2 \cdot 3p^2 = 15p^4$

$3k^2 \cdot 3k^3 = 9k^5$

$4p \cdot 3p^4 = 12p^5$

$4d^2 \cdot 3d^5 = 12d^7$

$5t^2 \cdot 5t = 25t^3$

$2q^4 \cdot 10q^3 = 20q^7$

$10t^6 \cdot 5t^2 = 50t^8$

$4n^2 \cdot 3n^2 = 12n^4$

$10h^6 \cdot 10h^2 = 100h^8$

$4v^5 \cdot 2v^6 = 8v^{11}$

Simplify each expression.

$5r^3 \cdot 2r^6 =$

$3w^6 \cdot 2w^5 =$

$10n^6 \cdot 3n^3 =$

$4y^3 \cdot 3y =$

$3g^2 \cdot 4g^5 =$

$3b \cdot 5b^4 =$

$10v^4 \cdot 4v^6 =$

$5v \cdot 3v^5 =$

$2r \cdot 3r^5 =$

$3x^2 \cdot 4x^5 =$

$10v^5 \cdot 5v^6 =$

$4v^2 \cdot 5v^6 =$

$2u^2 \cdot 5u^2 =$

$4y^5 \cdot 3y^6 =$

$5g^5 \cdot 10g^2 =$

$4g^4 \cdot 10g^2 =$

$2m \cdot 5m^3 =$

$5x^2 \cdot 2x^6 =$

$3p^2 \cdot 2p^4 =$

$5t^3 \cdot 4t^5 =$

# Answer Key

Simplify each expression.

$5r^3 \cdot 2r^6 = 10r^9$

$3w^6 \cdot 2w^5 = 6w^{11}$

$10n^6 \cdot 3n^3 = 30n^9$

$4y^3 \cdot 3y = 12y^4$

$3g^2 \cdot 4g^5 = 12g^7$

$3b \cdot 5b^4 = 15b^5$

$10v^4 \cdot 4v^6 = 40v^{10}$

$5v \cdot 3v^5 = 15v^6$

$2r \cdot 3r^5 = 6r^6$

$3x^2 \cdot 4x^5 = 12x^7$

$10v^5 \cdot 5v^6 = 50v^{11}$

$4v^2 \cdot 5v^6 = 20v^8$

$2u^2 \cdot 5u^2 = 10u^4$

$4y^5 \cdot 3y^6 = 12y^{11}$

$5g^5 \cdot 10g^2 = 50g^7$

$4g^4 \cdot 10g^2 = 40g^6$

$2m \cdot 5m^3 = 10m^4$

$5x^2 \cdot 2x^6 = 10x^8$

$3p^2 \cdot 2p^4 = 6p^6$

$5t^3 \cdot 4t^5 = 20t^8$

Simplify each expression.

$2q^4 \cdot 3q^3 =$

$2d^2 \cdot 3d^2 =$

$3d^3 \cdot 4d =$

$2t^3 \cdot 2t^5 =$

$2a \cdot 3a^2 =$

$2q^3 \cdot 10q^4 =$

$10f \cdot 3f^4 =$

$3g^6 \cdot 5g^3 =$

$2h^5 \cdot 3h^5 =$

$4v^5 \cdot 10v^6 =$

$5k^5 \cdot 4k =$

$4b^3 \cdot 10b^4 =$

$3r^4 \cdot 5r^5 =$

$3q^2 \cdot 10q =$

$10k^4 \cdot 2k =$

$3u^2 \cdot 3u =$

$4k^5 \cdot 2k =$

$10k^4 \cdot 5k^2 =$

$4w^3 \cdot 10w^2 =$

$2m^3 \cdot 5m =$

# Answer Key

Simplify each expression.

$2q^4 \cdot 3q^3 = 6q^7$

$2d^2 \cdot 3d^2 = 6d^4$

$3d^3 \cdot 4d = 12d^4$

$2t^3 \cdot 2t^5 = 4t^8$

$2a \cdot 3a^2 = 6a^3$

$2q^3 \cdot 10q^4 = 20q^7$

$10f \cdot 3f^4 = 30f^5$

$3g^6 \cdot 5g^3 = 15g^9$

$2h^5 \cdot 3h^5 = 6h^{10}$

$4v^5 \cdot 10v^6 = 40v^{11}$

$5k^5 \cdot 4k = 20k^6$

$4b^3 \cdot 10b^4 = 40b^7$

$3r^4 \cdot 5r^5 = 15r^9$

$3q^2 \cdot 10q = 30q^3$

$10k^4 \cdot 2k = 20k^5$

$3u^2 \cdot 3u = 9u^3$

$4k^5 \cdot 2k = 8k^6$

$10k^4 \cdot 5k^2 = 50k^6$

$4w^3 \cdot 10w^2 = 40w^5$

$2m^3 \cdot 5m = 10m^4$

Simplify each expression.

$10a^3 \cdot 2a^4 =$

$2k^3 \cdot 4k^5 =$

$3a^3 \cdot 4a^2 =$

$3t^4 \cdot 5t^4 =$

$5f^2 \cdot 5f =$

$4v \cdot 10v^4 =$

$3y \cdot 5y^2 =$

$2f^2 \cdot 3f^5 =$

$10k \cdot 3k^3 =$

$2a^4 \cdot 10a^6 =$

$4a^4 \cdot 10a^2 =$

$2f^4 \cdot 3f^5 =$

$3a \cdot 4a^3 =$

$4r \cdot 5r^6 =$

$2g^5 \cdot 10g^2 =$

$4k^2 \cdot 10k^2 =$

$3w^5 \cdot 2w^3 =$

$4k \cdot 3k^3 =$

$3x \cdot 10x^2 =$

$5b^2 \cdot 10b^4 =$

# Answer Key

Simplify each expression.

$10a^3 \cdot 2a^4 = 20a^7$

$2k^3 \cdot 4k^5 = 8k^8$

$3a^3 \cdot 4a^2 = 12a^5$

$3t^4 \cdot 5t^4 = 15t^8$

$5f^2 \cdot 5f = 25f^3$

$4v \cdot 10v^4 = 40v^5$

$3y \cdot 5y^2 = 15y^3$

$2f^2 \cdot 3f^5 = 6f^7$

$10k \cdot 3k^3 = 30k^4$

$2a^4 \cdot 10a^6 = 20a^{10}$

$4a^4 \cdot 10a^2 = 40a^6$

$2f^4 \cdot 3f^5 = 6f^9$

$3a \cdot 4a^3 = 12a^4$

$4r \cdot 5r^6 = 20r^7$

$2g^5 \cdot 10g^2 = 20g^7$

$4k^2 \cdot 10k^2 = 40k^4$

$3w^5 \cdot 2w^3 = 6w^8$

$4k \cdot 3k^3 = 12k^4$

$3x \cdot 10x^2 = 30x^3$

$5b^2 \cdot 10b^4 = 50b^6$

Simplify each expression.

$4b^5 \cdot 3b^3 =$

$10r^3 \cdot 2r^4 =$

$4y^4 \cdot 3y =$

$10a \cdot 5a^3 =$

$4q^3 \cdot 2q^3 =$

$4g \cdot 10g^3 =$

$3x \cdot 4x^2 =$

$3x^6 \cdot 4x^2 =$

$3a^3 \cdot 4a^4 =$

$3a \cdot 4a^2 =$

$10h^3 \cdot 3h^6 =$

$5h^5 \cdot 3h =$

$10x^2 \cdot 3x^5 =$

$5t \cdot 5t^3 =$

$10a^6 \cdot 3a^6 =$

$4m^2 \cdot 2m^3 =$

$4r^3 \cdot 4r^3 =$

$10v^4 \cdot 4v^5 =$

$3t^5 \cdot 5t =$

$3p^2 \cdot 10p^4 =$

# Answer Key

Simplify each expression.

$4b^5 \cdot 3b^3 = 12b^8$

$10r^3 \cdot 2r^4 = 20r^7$

$4y^4 \cdot 3y = 12y^5$

$10a \cdot 5a^3 = 50a^4$

$4q^3 \cdot 2q^3 = 8q^6$

$4g \cdot 10g^3 = 40g^4$

$3x \cdot 4x^2 = 12x^3$

$3x^6 \cdot 4x^2 = 12x^8$

$3a^3 \cdot 4a^4 = 12a^7$

$3a \cdot 4a^2 = 12a^3$

$10h^3 \cdot 3h^6 = 30h^9$

$5h^5 \cdot 3h = 15h^6$

$10x^2 \cdot 3x^5 = 30x^7$

$5t \cdot 5t^3 = 25t^4$

$10a^6 \cdot 3a^6 = 30a^{12}$

$4m^2 \cdot 2m^3 = 8m^5$

$4r^3 \cdot 4r^3 = 16r^6$

$10v^4 \cdot 4v^5 = 40v^9$

$3t^5 \cdot 5t = 15t^6$

$3p^2 \cdot 10p^4 = 30p^6$

Simplify each expression.

$5v^3 \cdot 4v =$

$10h \cdot 2h^2 =$

$3q^4 \cdot 10q^4 =$

$4k \cdot 4k^5 =$

$5h^5 \cdot 10h^3 =$

$5k^4 \cdot 3k^2 =$

$4u^6 \cdot 3u^6 =$

$3d^4 \cdot 3d^6 =$

$10m^3 \cdot 3m^6 =$

$2w \cdot 4w^4 =$

$10b \cdot 5b^5 =$

$3p^3 \cdot 5p^3 =$

$4b^4 \cdot 2b^3 =$

$5k^3 \cdot 3k^2 =$

$3f^6 \cdot 2f^5 =$

$2q^2 \cdot 2q^5 =$

$4q^6 \cdot 10q^2 =$

$10v^6 \cdot 5v^6 =$

$2d^5 \cdot 3d^3 =$

$3k^2 \cdot 10k^2 =$

# Answer Key

Simplify each expression.

$5v^3 \cdot 4v = 20v^4$

$10h \cdot 2h^2 = 20h^3$

$3q^4 \cdot 10q^4 = 30q^8$

$4k \cdot 4k^5 = 16k^6$

$5h^5 \cdot 10h^3 = 50h^8$

$5k^4 \cdot 3k^2 = 15k^6$

$4u^6 \cdot 3u^6 = 12u^{12}$

$3d^4 \cdot 3d^6 = 9d^{10}$

$10m^3 \cdot 3m^6 = 30m^9$

$2w \cdot 4w^4 = 8w^5$

$10b \cdot 5b^5 = 50b^6$

$3p^3 \cdot 5p^3 = 15p^6$

$4b^4 \cdot 2b^3 = 8b^7$

$5k^3 \cdot 3k^2 = 15k^5$

$3f^6 \cdot 2f^5 = 6f^{11}$

$2q^2 \cdot 2q^5 = 4q^7$

$4q^6 \cdot 10q^2 = 40q^8$

$10v^6 \cdot 5v^6 = 50v^{12}$

$2d^5 \cdot 3d^3 = 6d^8$

$3k^2 \cdot 10k^2 = 30k^4$

Simplify each expression.

$3f^6 \cdot 4f^4 =$

$2x^3 \cdot 5x^6 =$

$10m^2 \cdot 3m^5 =$

$5y^3 \cdot 5y^2 =$

$10f^5 \cdot 3f^3 =$

$10v^2 \cdot 10v =$

$10g^2 \cdot 5g^2 =$

$4m^3 \cdot 3m^3 =$

$4q^5 \cdot 2q^4 =$

$10a^3 \cdot 4a^6 =$

$10b^3 \cdot 2b^3 =$

$4n \cdot 3n^4 =$

$4h^6 \cdot 2h^6 =$

$5u \cdot 3u^5 =$

$4n^2 \cdot 2n^3 =$

$2t^5 \cdot 4t^4 =$

$4q^6 \cdot 4q^3 =$

$2k^5 \cdot 5k =$

$10t^3 \cdot 3t =$

$4x^2 \cdot 10x^3 =$

# Answer Key

Simplify each expression.

$3f^6 \cdot 4f^4 = 12f^{10}$

$2x^3 \cdot 5x^6 = 10x^9$

$10m^2 \cdot 3m^5 = 30m^7$

$5y^3 \cdot 5y^2 = 25y^5$

$10f^5 \cdot 3f^3 = 30f^8$

$10v^2 \cdot 10v = 100v^3$

$10g^2 \cdot 5g^2 = 50g^4$

$4m^3 \cdot 3m^3 = 12m^6$

$4q^5 \cdot 2q^4 = 8q^9$

$10a^3 \cdot 4a^6 = 40a^9$

$10b^3 \cdot 2b^3 = 20b^6$

$4n \cdot 3n^4 = 12n^5$

$4h^6 \cdot 2h^6 = 8h^{12}$

$5u \cdot 3u^5 = 15u^6$

$4n^2 \cdot 2n^3 = 8n^5$

$2t^5 \cdot 4t^4 = 8t^9$

$4q^6 \cdot 4q^3 = 16q^9$

$2k^5 \cdot 5k = 10k^6$

$10t^3 \cdot 3t = 30t^4$

$4x^2 \cdot 10x^3 = 40x^5$

Simplify each expression.

$4v^5 \cdot 5v =$

$3n^6 \cdot 3n^5 =$

$5n^4 \cdot 3n =$

$4f^2 \cdot 5f^3 =$

$5a^3 \cdot 2a^2 =$

$5r^4 \cdot 10r^6 =$

$4h^4 \cdot 3h^6 =$

$3u^2 \cdot 5u^2 =$

$3p^4 \cdot 4p =$

$4h^2 \cdot 4h =$

$10t^3 \cdot 5t^4 =$

$3q \cdot 4q^2 =$

$10f \cdot 2f^4 =$

$2q^3 \cdot 2q^6 =$

$10h^3 \cdot 2h^4 =$

$10q^3 \cdot 2q^4 =$

$3d^2 \cdot 10d =$

$2h^2 \cdot 10h^4 =$

$10k \cdot 3k^2 =$

$5a^2 \cdot 2a^5 =$

# Answer Key

Simplify each expression.

$4v^5 \cdot 5v = 20v^6$

$3n^6 \cdot 3n^5 = 9n^{11}$

$5n^4 \cdot 3n = 15n^5$

$4f^2 \cdot 5f^3 = 20f^5$

$5a^3 \cdot 2a^2 = 10a^5$

$5r^4 \cdot 10r^6 = 50r^{10}$

$4h^4 \cdot 3h^6 = 12h^{10}$

$3u^2 \cdot 5u^2 = 15u^4$

$3p^4 \cdot 4p = 12p^5$

$4h^2 \cdot 4h = 16h^3$

$10t^3 \cdot 5t^4 = 50t^7$

$3q \cdot 4q^2 = 12q^3$

$10f \cdot 2f^4 = 20f^5$

$2q^3 \cdot 2q^6 = 4q^9$

$10h^3 \cdot 2h^4 = 20h^7$

$10q^3 \cdot 2q^4 = 20q^7$

$3d^2 \cdot 10d = 30d^3$

$2h^2 \cdot 10h^4 = 20h^6$

$10k \cdot 3k^2 = 30k^3$

$5a^2 \cdot 2a^5 = 10a^7$

Simplify each expression.

$2h^3g^3 \cdot 3h^3g^3 =$

$4m^5q^4 \cdot 4m^5q^6 =$

$4r^6k^2 \cdot 3r^5k^5 =$

$4q^6f^2 \cdot 5q^6f^2 =$

$5m^6q^6 \cdot 3mq^5 =$

$10a^6b^3 \cdot 10a^6b^4 =$

$2bk^5 \cdot 10b^3k^4 =$

$10x^6h \cdot 2x^3h^6 =$

$10g^5m \cdot 2g^2m^2 =$

$5u^5b^2 \cdot 3u^5b^4 =$

$2u^2a^5 \cdot 10ua^5 =$

$4m^3x^2 \cdot 4m^3x^2 =$

$5k^4q^5 \cdot 2k^2q^2 =$

$10k^5v^3 \cdot 3k^3v^3 =$

$3t^2a \cdot 10t^5a^6 =$

$3ga^6 \cdot 4g^2a^2 =$

$2q^5m \cdot 5q^3m^2 =$

$2k^2x^2 \cdot 3k^2x =$

$5g^3u^4 \cdot 5g^5u^5 =$

$10y^6p^6 \cdot 4y^3p^3 =$

# Answer Key

Simplify each expression.

$2h^3g^3 \cdot 3h^3g^3 = 6h^6g^6$

$4r^6k^2 \cdot 3r^5k^5 = 12r^{11}k^7$

$5m^6q^6 \cdot 3mq^5 = 15m^7q^{11}$

$2bk^5 \cdot 10b^3k^4 = 20b^4k^9$

$10g^5m \cdot 2g^2m^2 = 20g^7m^3$

$2u^2a^5 \cdot 10ua^5 = 20u^3a^{10}$

$5k^4q^5 \cdot 2k^2q^2 = 10k^6q^7$

$3t^2a \cdot 10t^5a^6 = 30t^7a^7$

$2q^5m \cdot 5q^3m^2 = 10q^8m^3$

$5g^3u^4 \cdot 5g^5u^5 = 25g^8u^9$

$4m^5q^4 \cdot 4m^5q^6 = 16m^{10}q^{10}$

$4q^6f^2 \cdot 5q^6f^2 = 20q^{12}f^4$

$10a^6b^3 \cdot 10a^6b^4 = 100a^{12}b^7$

$10x^6h \cdot 2x^3h^6 = 20x^9h^7$

$5u^5b^2 \cdot 3u^5b^4 = 15u^{10}b^6$

$4m^3x^2 \cdot 4m^3x^2 = 16m^6x^4$

$10k^5v^3 \cdot 3k^3v^3 = 30k^8v^6$

$3ga^6 \cdot 4g^2a^2 = 12g^3a^8$

$2k^2x^2 \cdot 3k^2x = 6k^4x^3$

$10y^6p^6 \cdot 4y^3p^3 = 40y^9p^9$

Simplify each expression.

$3u^2a^6 \cdot 4u^3a^3 =$

$3w^3a^2 \cdot 5w^2a^2 =$

$10a^6q^4 \cdot 10a^6q^4 =$

$10p^2d^2 \cdot 2p^3d =$

$5x^2b^4 \cdot 3x^2b^6 =$

$10h^5d^2 \cdot 4h^4d^2 =$

$3w^5f^5 \cdot 10w^2f^5 =$

$2u^5v^2 \cdot 3u^2v =$

$4u^5y^6 \cdot 2uy^6 =$

$2bw^4 \cdot 2b^5w^6 =$

$4m^2d^4 \cdot 10m^5d^6 =$

$2k^6t^2 \cdot 5kt^2 =$

$5rd^2 \cdot 4r^6d^4 =$

$2r^6q^2 \cdot 5r^4q^3 =$

$3a^5h^2 \cdot 2ah =$

$10t^3m^4 \cdot 10tm =$

$5y^6x^6 \cdot 4y^3x^5 =$

$2r^2g^5 \cdot 10r^2g^5 =$

$10d^6h^2 \cdot 2d^2h^3 =$

$2rk^5 \cdot 10r^5k^2 =$

# Answer Key

Simplify each expression.

$3u^2a^6 \cdot 4u^3a^3 = 12u^5a^9$

$3w^3a^2 \cdot 5w^2a^2 = 15w^5a^4$

$10a^6q^4 \cdot 10a^6q^4 = 100a^{12}q^8$

$10p^2d^2 \cdot 2p^3d = 20p^5d^3$

$5x^2b^4 \cdot 3x^2b^6 = 15x^4b^{10}$

$10h^5d^2 \cdot 4h^4d^2 = 40h^9d^4$

$3w^5f^5 \cdot 10w^2f^5 = 30w^7f^{10}$

$2u^5v^2 \cdot 3u^2v = 6u^7v^3$

$4u^5y^6 \cdot 2uy^6 = 8u^6y^{12}$

$2bw^4 \cdot 2b^5w^6 = 4b^6w^{10}$

$4m^2d^4 \cdot 10m^5d^6 = 40m^7d^{10}$

$2k^6t^2 \cdot 5kt^2 = 10k^7t^4$

$5rd^2 \cdot 4r^6d^4 = 20r^7d^6$

$2r^6q^2 \cdot 5r^4q^3 = 10r^{10}q^5$

$3a^5h^2 \cdot 2ah = 6a^6h^3$

$10t^3m^4 \cdot 10tm = 100t^4m^5$

$5y^6x^6 \cdot 4y^3x^5 = 20y^9x^{11}$

$2r^2g^5 \cdot 10r^2g^5 = 20r^4g^{10}$

$10d^6h^2 \cdot 2d^2h^3 = 20d^8h^5$

$2rk^5 \cdot 10r^5k^2 = 20r^6k^7$

Simplify each expression.

$3f^4p^5 \cdot 2f^2p^3 =$

$3f^3x^2 \cdot 4f^3x^4 =$

$3y^5r^6 \cdot 5yr =$

$3t^2h^3 \cdot 5t^3h^3 =$

$3h^6w^6 \cdot 4h^3w^6 =$

$3g^4d^2 \cdot 3g^5d =$

$5xb \cdot 5x^3b^2 =$

$2x^4u^5 \cdot 10x^4u^3 =$

$4g^3p^4 \cdot 4gp^2 =$

$2k^4v^3 \cdot 2k^6v^6 =$

$10d^2a^3 \cdot 2d^5a^3 =$

$5a^5d \cdot 3a^2d^5 =$

$5q^4p^3 \cdot 5q^3p^4 =$

$5fr \cdot 2f^2r^2 =$

$3w^2b^3 \cdot 5w^4b^5 =$

$10r^5u^3 \cdot 4r^4u^5 =$

$3mb^4 \cdot 10mb^6 =$

$5u^3r^3 \cdot 10u^4r^6 =$

$5v^4b^5 \cdot 4vb^3 =$

$4q^4g^5 \cdot 3qg^2 =$

# Answer Key

Simplify each expression.

$3f^4p^5 \cdot 2f^2p^3 = 6f^6p^8$

$10d^2a^3 \cdot 2d^5a^3 = 20d^7a^6$

$3f^3x^2 \cdot 4f^3x^4 = 12f^6x^6$

$5a^5d \cdot 3a^2d^5 = 15a^7d^6$

$3y^5r^6 \cdot 5yr = 15y^6r^7$

$5q^4p^3 \cdot 5q^3p^4 = 25q^7p^7$

$3t^2h^3 \cdot 5t^3h^3 = 15t^5h^6$

$5fr \cdot 2f^2r^2 = 10f^3r^3$

$3h^6w^6 \cdot 4h^3w^6 = 12h^9w^{12}$

$3w^2b^3 \cdot 5w^4b^5 = 15w^6b^8$

$3g^4d^2 \cdot 3g^5d = 9g^9d^3$

$10r^5u^3 \cdot 4r^4u^5 = 40r^9u^8$

$5xb \cdot 5x^3b^2 = 25x^4b^3$

$3mb^4 \cdot 10mb^6 = 30m^2b^{10}$

$2x^4u^5 \cdot 10x^4u^3 = 20x^8u^8$

$5u^3r^3 \cdot 10u^4r^6 = 50u^7r^9$

$4g^3p^4 \cdot 4gp^2 = 16g^4p^6$

$5v^4b^5 \cdot 4vb^3 = 20v^5b^8$

$2k^4v^3 \cdot 2k^6v^6 = 4k^{10}v^9$

$4q^4g^5 \cdot 3qg^2 = 12q^5g^7$

Simplify each expression.

$3y^2r^3 \cdot 3y^5r^4 =$

$2u^4g^6 \cdot 5u^4g^4 =$

$2a^3w^6 \cdot 4a^2w^2 =$

$3t^2v^4 \cdot 3t^3v^5 =$

$10n^3d^4 \cdot 10n^2d^3 =$

$4r^3g^4 \cdot 5r^3g =$

$10kg^5 \cdot 5kg^2 =$

$5d^2r^3 \cdot 5d^5r^3 =$

$4tu \cdot 3tu^2 =$

$4f^4g^3 \cdot 3f^2g =$

$4k^5v \cdot 10k^5v^2 =$

$3m^6f^5 \cdot 2m^3f^5 =$

$4n^5d^5 \cdot 4n^6d^5 =$

$3v^2n^3 \cdot 2v^6n^3 =$

$4yu \cdot 2y^3u^2 =$

$4t^5x^3 \cdot 4t^3x^5 =$

$4g^2v^5 \cdot 5g^3v^4 =$

$4n^2y^2 \cdot 5n^3y^4 =$

$3w^4b^4 \cdot 4w^4b^5 =$

$10a^3d^5 \cdot 4a^4d^5 =$

# Answer Key

Simplify each expression.

$3y^2r^3 \cdot 3y^5r^4 = 9y^7r^7$

$2a^3w^6 \cdot 4a^2w^2 = 8a^5w^8$

$10n^3d^4 \cdot 10n^2d^3 = 100n^5d^7$

$10kg^5 \cdot 5kg^2 = 50k^2g^7$

$4tu \cdot 3tu^2 = 12t^2u^3$

$4k^5v \cdot 10k^5v^2 = 40k^{10}v^3$

$4n^5d^5 \cdot 4n^6d^5 = 16n^{11}d^{10}$

$4yu \cdot 2y^3u^2 = 8y^4u^3$

$4g^2v^5 \cdot 5g^3v^4 = 20g^5v^9$

$3w^4b^4 \cdot 4w^4b^5 = 12w^8b^9$

$2u^4g^6 \cdot 5u^4g^4 = 10u^8g^{10}$

$3t^2v^4 \cdot 3t^3v^5 = 9t^5v^9$

$4r^3g^4 \cdot 5r^3g = 20r^6g^5$

$5d^2r^3 \cdot 5d^5r^3 = 25d^7r^6$

$4f^4g^3 \cdot 3f^2g = 12f^6g^4$

$3m^6f^5 \cdot 2m^3f^5 = 6m^9f^{10}$

$3v^2n^3 \cdot 2v^6n^3 = 6v^8n^6$

$4t^5x^3 \cdot 4t^3x^5 = 16t^8x^8$

$4n^2y^2 \cdot 5n^3y^4 = 20n^5y^6$

$10a^3d^5 \cdot 4a^4d^5 = 40a^7d^{10}$

Simplify each expression.

$3b^4v^2 \cdot 3b^4v =$

$2r^4g^6 \cdot 2r^4g =$

$5h^3a^2 \cdot 2h^6a^2 =$

$10w^2d^3 \cdot 3w^4d^2 =$

$10p^3m^2 \cdot 2p^3m^3 =$

$2g^6n^3 \cdot 4g^5n^6 =$

$3r^2p^5 \cdot 3r^5p^6 =$

$3n^2p \cdot 3np^5 =$

$3b^6f^2 \cdot 3bf^4 =$

$4r^3a^5 \cdot 4r^4a^2 =$

$3a^4w^2 \cdot 4a^4w =$

$4hb^2 \cdot 10h^2b^4 =$

$10yh^2 \cdot 4y^2h^4 =$

$10a^6k^5 \cdot 3a^4k^3 =$

$2db^2 \cdot 2d^4b^6 =$

$5a^3q^3 \cdot 3a^2q^4 =$

$10u^4v \cdot 5u^2v^2 =$

$3v^4u^5 \cdot 2v^3u^3 =$

$4n^4d^5 \cdot 4n^4d^6 =$

$3ag^2 \cdot 3a^4g^3 =$

# Answer Key

Simplify each expression.

$3b^4v^2 \cdot 3b^4v = 9b^8v^3$

$2r^4g^6 \cdot 2r^4g = 4r^8g^7$

$5h^3a^2 \cdot 2h^6a^2 = 10h^9a^4$

$10w^2d^3 \cdot 3w^4d^2 = 30w^6d^5$

$10p^3m^2 \cdot 2p^3m^3 = 20p^6m^5$

$2g^6n^3 \cdot 4g^5n^6 = 8g^{11}n^9$

$3r^2p^5 \cdot 3r^5p^6 = 9r^7p^{11}$

$3n^2p \cdot 3np^5 = 9n^3p^6$

$3b^6f^2 \cdot 3bf^4 = 9b^7f^6$

$4r^3a^5 \cdot 4r^4a^2 = 16r^7a^7$

$3a^4w^2 \cdot 4a^4w = 12a^8w^3$

$4hb^2 \cdot 10h^2b^4 = 40h^3b^6$

$10yh^2 \cdot 4y^2h^4 = 40y^3h^6$

$10a^6k^5 \cdot 3a^4k^3 = 30a^{10}k^8$

$2db^2 \cdot 2d^4b^6 = 4d^5b^8$

$5a^3q^3 \cdot 3a^2q^4 = 15a^5q^7$

$10u^4v \cdot 5u^2v^2 = 50u^6v^3$

$3v^4u^5 \cdot 2v^3u^3 = 6v^7u^8$

$4n^4d^5 \cdot 4n^4d^6 = 16n^8d^{11}$

$3ag^2 \cdot 3a^4g^3 = 9a^5g^5$

Simplify each expression.

$3a^4m^5 \cdot 2a^6m^4 =$

$10h^6p^5 \cdot 10h^4p =$

$4b^5w^6 \cdot 4b^5w^4 =$

$3u^4k \cdot 3uk^4 =$

$4n^4r \cdot 3n^2r^5 =$

$5n^5x^3 \cdot 10nx^6 =$

$5r^4g^5 \cdot 2r^5g^3 =$

$5q^3d \cdot 5q^3d =$

$2d^2a^4 \cdot 5d^3a^5 =$

$5b^3p^3 \cdot 4b^3p^2 =$

$2p^5q \cdot 4p^5q^4 =$

$4g^3q^2 \cdot 4gq^3 =$

$3w^4g^2 \cdot 5wg^4 =$

$3k^3x \cdot 3k^5x^4 =$

$5qb^3 \cdot 3q^6b^5 =$

$2t^3q^2 \cdot 5tq^2 =$

$2b^4k^5 \cdot 10b^6k^2 =$

$4um^5 \cdot 3u^4m^2 =$

$3f^5m^4 \cdot 2f^3m^2 =$

$2a^6d^2 \cdot 10a^5d =$

# Answer Key

Simplify each expression.

$3a^4m^5 \cdot 2a^6m^4 = 6a^{10}m^9$

$10h^6p^5 \cdot 10h^4p = 100h^{10}p^6$

$4b^5w^6 \cdot 4b^5w^4 = 16b^{10}w^{10}$

$3u^4k \cdot 3uk^4 = 9u^5k^5$

$4n^4r \cdot 3n^2r^5 = 12n^6r^6$

$5n^5x^3 \cdot 10nx^6 = 50n^6x^9$

$5r^4g^5 \cdot 2r^5g^3 = 10r^9g^8$

$5q^3d \cdot 5q^3d = 25q^6d^2$

$2d^2a^4 \cdot 5d^3a^5 = 10d^5a^9$

$5b^3p^3 \cdot 4b^3p^2 = 20b^6p^5$

$2p^5q \cdot 4p^5q^4 = 8p^{10}q^5$

$4g^3q^2 \cdot 4gq^3 = 16g^4q^5$

$3w^4g^2 \cdot 5wg^4 = 15w^5g^6$

$3k^3x \cdot 3k^5x^4 = 9k^8x^5$

$5qb^3 \cdot 3q^6b^5 = 15q^7b^8$

$2t^3q^2 \cdot 5tq^2 = 10t^4q^4$

$2b^4k^5 \cdot 10b^6k^2 = 20b^{10}k^7$

$4um^5 \cdot 3u^4m^2 = 12u^5m^7$

$3f^5m^4 \cdot 2f^3m^2 = 6f^8m^6$

$2a^6d^2 \cdot 10a^5d = 20a^{11}d^3$

Simplify each expression.

$2d^2f \cdot 2d^2f^4 =$

$5b^4x^5 \cdot 4bx^5 =$

$4mv^2 \cdot 4m^4v^4 =$

$2r^4g^2 \cdot 2rg^5 =$

$4u^4f^5 \cdot 3u^6f^5 =$

$5t^3d^4 \cdot 5t^4d^4 =$

$3dt^6 \cdot 10d^3t^4 =$

$5p^6y^2 \cdot 5py^5 =$

$4x^3q^2 \cdot 4x^2q^3 =$

$3hw \cdot 10h^4w^3 =$

$2g^3u \cdot 3g^4u^3 =$

$3a^3g^4 \cdot 5a^2g =$

$10d^3h^3 \cdot 3dh^3 =$

$2d^6x \cdot 10dx =$

$5w^3h^3 \cdot 4w^2h^5 =$

$2t^4p^6 \cdot 2t^3p^2 =$

$3v^6t^6 \cdot 3v^2t^2 =$

$2qm^4 \cdot 4q^6m =$

$10q^3g^4 \cdot 4q^2g^4 =$

$2n^3q^3 \cdot 4n^2q^4 =$

# Answer Key

Simplify each expression.

$2d^2f \cdot 2d^2f^4 = 4d^4f^5$

$5b^4x^5 \cdot 4bx^5 = 20b^5x^{10}$

$4mv^2 \cdot 4m^4v^4 = 16m^5v^6$

$2r^4g^2 \cdot 2rg^5 = 4r^5g^7$

$4u^4f^5 \cdot 3u^6f^5 = 12u^{10}f^{10}$

$5t^3d^4 \cdot 5t^4d^4 = 25t^7d^8$

$3dt^6 \cdot 10d^3t^4 = 30d^4t^{10}$

$5p^6y^2 \cdot 5py^5 = 25p^7y^7$

$4x^3q^2 \cdot 4x^2q^3 = 16x^5q^5$

$3hw \cdot 10h^4w^3 = 30h^5w^4$

$2g^3u \cdot 3g^4u^3 = 6g^7u^4$

$3a^3g^4 \cdot 5a^2g = 15a^5g^5$

$10d^3h^3 \cdot 3dh^3 = 30d^4h^6$

$2d^6x \cdot 10dx = 20d^7x^2$

$5w^3h^3 \cdot 4w^2h^5 = 20w^5h^8$

$2t^4p^6 \cdot 2t^3p^2 = 4t^7p^8$

$3v^6t^6 \cdot 3v^2t^2 = 9v^8t^8$

$2qm^4 \cdot 4q^6m = 8q^7m^5$

$10q^3g^4 \cdot 4q^2g^4 = 40q^5g^8$

$2n^3q^3 \cdot 4n^2q^4 = 8n^5q^7$

Simplify each expression.

$5b^6f^2 \cdot 3b^4f^3 =$

$4w^2g \cdot 5w^2g^2 =$

$3m^2h^4 \cdot 3m^5h^2 =$

$5y^5r \cdot 3y^5r =$

$10dv \cdot 3d^2v =$

$3t^5r \cdot 4t^2r =$

$4x^3b^5 \cdot 5x^2b^5 =$

$5m^2w^2 \cdot 4m^2w =$

$2b^5t^2 \cdot 4bt^3 =$

$3m^4b^2 \cdot 2mb^2 =$

$10fk^5 \cdot 4f^6k^5 =$

$10r^5p^4 \cdot 3r^3p^2 =$

$4u^6h^5 \cdot 10u^5h =$

$4m^5p^4 \cdot 3m^5p^6 =$

$3am \cdot 5a^3m^3 =$

$5b^5m \cdot 2b^4m^4 =$

$4r^2q \cdot 2r^5q^2 =$

$4m^6f^6 \cdot 5m^6f^3 =$

$5v^6k^4 \cdot 3v^3k^5 =$

$5f^3w^5 \cdot 10fw^3 =$

# Answer Key

Simplify each expression.

$5b^6f^2 \cdot 3b^4f^3 = 15b^{10}f^5$

$4w^2g \cdot 5w^2g^2 = 20w^4g^3$

$3m^2h^4 \cdot 3m^5h^2 = 9m^7h^6$

$5y^5r \cdot 3y^5r = 15y^{10}r^2$

$10dv \cdot 3d^2v = 30d^3v^2$

$3t^5r \cdot 4t^2r = 12t^7r^2$

$4x^3b^5 \cdot 5x^2b^5 = 20x^5b^{10}$

$5m^2w^2 \cdot 4m^2w = 20m^4w^3$

$2b^5t^2 \cdot 4bt^3 = 8b^6t^5$

$3m^4b^2 \cdot 2mb^2 = 6m^5b^4$

$10fk^5 \cdot 4f^6k^5 = 40f^7k^{10}$

$10r^5p^4 \cdot 3r^3p^2 = 30r^8p^6$

$4u^6h^5 \cdot 10u^5h = 40u^{11}h^6$

$4m^5p^4 \cdot 3m^5p^6 = 12m^{10}p^{10}$

$3am \cdot 5a^3m^3 = 15a^4m^4$

$5b^5m \cdot 2b^4m^4 = 10b^9m^5$

$4r^2q \cdot 2r^5q^2 = 8r^7q^3$

$4m^6f^6 \cdot 5m^6f^3 = 20m^{12}f^9$

$5v^6k^4 \cdot 3v^3k^5 = 15v^9k^9$

$5f^3w^5 \cdot 10fw^3 = 50f^4w^8$

Simplify each expression.

$4h^2t^5 \cdot 5h^2t^4 =$

$2w^6p^3 \cdot 5w^2p^5 =$

$3a^2f^6 \cdot 3a^6f^3 =$

$2yd^5 \cdot 5y^4d^6 =$

$4g^4n^5 \cdot 2g^5n^2 =$

$5u^2t \cdot 3ut^3 =$

$2w^4h^6 \cdot 10w^4h^2 =$

$4m^6w^3 \cdot 5m^4w^3 =$

$10qy^2 \cdot 3q^2y^2 =$

$4y^3q^3 \cdot 5y^6q =$

$10r^5h^2 \cdot 5r^4h^5 =$

$10uv^6 \cdot 4u^6v^5 =$

$10n^4x^2 \cdot 3n^4x^4 =$

$2p^2f^6 \cdot 10pf^2 =$

$2pm \cdot 4p^3m^2 =$

$3a^5b^4 \cdot 4a^5b^4 =$

$3b^6k^3 \cdot 3b^6k =$

$10h^3y^6 \cdot 3h^4y^4 =$

$2yd^6 \cdot 2y^2d^5 =$

$3w^6x \cdot 2wx =$

# Answer Key

Simplify each expression.

$4h^2t^5 \cdot 5h^2t^4 = 20h^4t^9$

$2w^6p^3 \cdot 5w^2p^5 = 10w^8p^8$

$3a^2f^6 \cdot 3a^6f^3 = 9a^8f^9$

$2yd^5 \cdot 5y^4d^6 = 10y^5d^{11}$

$4g^4n^5 \cdot 2g^5n^2 = 8g^9n^7$

$5u^2t \cdot 3ut^3 = 15u^3t^4$

$2w^4h^6 \cdot 10w^4h^2 = 20w^8h^8$

$4m^6w^3 \cdot 5m^4w^3 = 20m^{10}w^6$

$10qy^2 \cdot 3q^2y^2 = 30q^3y^4$

$4y^3q^3 \cdot 5y^6q = 20y^9q^4$

$10r^5h^2 \cdot 5r^4h^5 = 50r^9h^7$

$10uv^6 \cdot 4u^6v^5 = 40u^7v^{11}$

$10n^4x^2 \cdot 3n^4x^4 = 30n^8x^6$

$2p^2f^6 \cdot 10pf^2 = 20p^3f^8$

$2pm \cdot 4p^3m^2 = 8p^4m^3$

$3a^5b^4 \cdot 4a^5b^4 = 12a^{10}b^8$

$3b^6k^3 \cdot 3b^6k = 9b^{12}k^4$

$10h^3y^6 \cdot 3h^4y^4 = 30h^7y^{10}$

$2yd^6 \cdot 2y^2d^5 = 4y^3d^{11}$

$3w^6x \cdot 2wx = 6w^7x^2$

Simplify each expression.

$10q^2d \cdot 3q^3d^5 =$

$5hk^2 \cdot 3h^3k^5 =$

$5q^5x^5 \cdot 5q^6x^5 =$

$5xk^3 \cdot 4x^4k^4 =$

$4x^6q^4 \cdot 3x^4q^3 =$

$10h^6f \cdot 4h^5f^4 =$

$3na^5 \cdot 5n^6a^6 =$

$3q^4p^4 \cdot 10q^3p^3 =$

$3x^2q^4 \cdot 10x^3q^6 =$

$2h^2a^3 \cdot 2ha^4 =$

$3a^3k^5 \cdot 10ak =$

$2u^6f^2 \cdot 3uf^2 =$

$4n^2h^6 \cdot 3n^2h^5 =$

$5n^5t \cdot 2n^3t^6 =$

$2r^6x^3 \cdot 2rx^6 =$

$5hd^6 \cdot 3h^2d^2 =$

$4d^2y^4 \cdot 4d^6y^3 =$

$10fn^2 \cdot 2f^3n^3 =$

$3k^3u^2 \cdot 3ku^2 =$

$2f^4r^4 \cdot 3fr^6 =$

# Answer Key

Simplify each expression.

$10q^2d \cdot 3q^3d^5 = 30q^5d^6$

$5hk^2 \cdot 3h^3k^5 = 15h^4k^7$

$5q^5x^5 \cdot 5q^6x^5 = 25q^{11}x^{10}$

$5xk^3 \cdot 4x^4k^4 = 20x^5k^7$

$4x^6q^4 \cdot 3x^4q^3 = 12x^{10}q^7$

$10h^6f \cdot 4h^5f^4 = 40h^{11}f^5$

$3na^5 \cdot 5n^6a^6 = 15n^7a^{11}$

$3q^4p^4 \cdot 10q^3p^3 = 30q^7p^7$

$3x^2q^4 \cdot 10x^3q^6 = 30x^5q^{10}$

$2h^2a^3 \cdot 2ha^4 = 4h^3a^7$

$3a^3k^5 \cdot 10ak = 30a^4k^6$

$2u^6f^2 \cdot 3uf^2 = 6u^7f^4$

$4n^2h^6 \cdot 3n^2h^5 = 12n^4h^{11}$

$5n^5t \cdot 2n^3t^6 = 10n^8t^7$

$2r^6x^3 \cdot 2rx^6 = 4r^7x^9$

$5hd^6 \cdot 3h^2d^2 = 15h^3d^8$

$4d^2y^4 \cdot 4d^6y^3 = 16d^8y^7$

$10fn^2 \cdot 2f^3n^3 = 20f^4n^5$

$3k^3u^2 \cdot 3ku^2 = 9k^4u^4$

$2f^4r^4 \cdot 3fr^6 = 6f^5r^{10}$

Simplify each expression.

$2y^4h^2 \cdot 4y^5h^4 =$

$3q^5g \cdot 2q^5g^6 =$

$5p^3q^5 \cdot 2p^2q^4 =$

$2wa^4 \cdot 5w^4a =$

$2u^3t \cdot 5u^3t^4 =$

$3f^5t^5 \cdot 4ft^2 =$

$2h^3a \cdot 4h^4a^4 =$

$2a^4w^4 \cdot 4a^3w^3 =$

$3xp^5 \cdot 5xp^2 =$

$5d^3n^4 \cdot 3d^4n =$

$3ud \cdot 5u^5d^3 =$

$10r^2p^4 \cdot 5r^5p^4 =$

$3d^3t^6 \cdot 5d^2t^6 =$

$4p^5d \cdot 10p^3d^4 =$

$4p^5g^3 \cdot 3p^2g^4 =$

$3xq^4 \cdot 5x^5q^3 =$

$10up^5 \cdot 3u^6p^4 =$

$3u^3n^5 \cdot 5u^5n^3 =$

$3mb^3 \cdot 5m^2b^2 =$

$5nq^2 \cdot 5n^3q^4 =$

# Answer Key

Simplify each expression.

$2y^4h^2 \cdot 4y^5h^4 = 8y^9h^6$

$3q^5g \cdot 2q^5g^6 = 6q^{10}g^7$

$5p^3q^5 \cdot 2p^2q^4 = 10p^5q^9$

$2wa^4 \cdot 5w^4a = 10w^5a^5$

$2u^3t \cdot 5u^3t^4 = 10u^6t^5$

$3f^5t^5 \cdot 4ft^2 = 12f^6t^7$

$2h^3a \cdot 4h^4a^4 = 8h^7a^5$

$2a^4w^4 \cdot 4a^3w^3 = 8a^7w^7$

$3xp^5 \cdot 5xp^2 = 15x^2p^7$

$5d^3n^4 \cdot 3d^4n = 15d^7n^5$

$3ud \cdot 5u^5d^3 = 15u^6d^4$

$10r^2p^4 \cdot 5r^5p^4 = 50r^7p^8$

$3d^3t^6 \cdot 5d^2t^6 = 15d^5t^{12}$

$4p^5d \cdot 10p^3d^4 = 40p^8d^5$

$4p^5g^3 \cdot 3p^2g^4 = 12p^7g^7$

$3xq^4 \cdot 5x^5q^3 = 15x^6q^7$

$10up^5 \cdot 3u^6p^4 = 30u^7p^9$

$3u^3n^5 \cdot 5u^5n^3 = 15u^8n^8$

$3mb^3 \cdot 5m^2b^2 = 15m^3b^5$

$5nq^2 \cdot 5n^3q^4 = 25n^4q^6$

Simplify each expression.

$5k^3h^6 \cdot 2k^4h^4 =$

$10q^6k^3 \cdot 10q^4k^5 =$

$3p^6y^4 \cdot 5p^3y^6 =$

$10q^3h^2 \cdot 3q^6h^4 =$

$3g^3y^6 \cdot 3g^3y^4 =$

$5f^4m^2 \cdot 3f^5m^4 =$

$3mt \cdot 2m^3t^3 =$

$10m^5n^3 \cdot 5m^3n^3 =$

$3hu^2 \cdot 3h^5u^6 =$

$2n^2k^2 \cdot 4n^3k =$

$3a^4w^4 \cdot 4aw^6 =$

$2f^4k^4 \cdot 4fk^2 =$

$3p^2d^4 \cdot 2pd^3 =$

$4h^5n^6 \cdot 2h^2n^4 =$

$2a^3y^4 \cdot 3ay^4 =$

$3f^5p^2 \cdot 4fp^4 =$

$5h^5u^2 \cdot 4h^6u^6 =$

$5p^6h^2 \cdot 5ph^2 =$

$4u^2g^2 \cdot 4u^2g =$

$10f^4q^6 \cdot 5f^3q^2 =$

# Answer Key

Simplify each expression.

$5k^3h^6 \cdot 2k^4h^4 = 10k^7h^{10}$

$10q^6k^3 \cdot 10q^4k^5 = 100q^{10}k^8$

$3p^6y^4 \cdot 5p^3y^6 = 15p^9y^{10}$

$10q^3h^2 \cdot 3q^6h^4 = 30q^9h^6$

$3g^3y^6 \cdot 3g^3y^4 = 9g^6y^{10}$

$5f^4m^2 \cdot 3f^5m^4 = 15f^9m^6$

$3mt \cdot 2m^3t^3 = 6m^4t^4$

$10m^5n^3 \cdot 5m^3n^3 = 50m^8n^6$

$3hu^2 \cdot 3h^5u^6 = 9h^6u^8$

$2n^2k^2 \cdot 4n^3k = 8n^5k^3$

$3a^4w^4 \cdot 4aw^6 = 12a^5w^{10}$

$2f^4k^4 \cdot 4fk^2 = 8f^5k^6$

$3p^2d^4 \cdot 2pd^3 = 6p^3d^7$

$4h^5n^6 \cdot 2h^2n^4 = 8h^7n^{10}$

$2a^3y^4 \cdot 3ay^4 = 6a^4y^8$

$3f^5p^2 \cdot 4fp^4 = 12f^6p^6$

$5h^5u^2 \cdot 4h^6u^6 = 20h^{11}u^8$

$5p^6h^2 \cdot 5ph^2 = 25p^7h^4$

$4u^2g^2 \cdot 4u^2g = 16u^4g^3$

$10f^4q^6 \cdot 5f^3q^2 = 50f^7q^8$

Simplify each expression.

$2bf^3 \cdot 2b^4f =$

$3k^5m^3 \cdot 2k^3m^5 =$

$4nx^3 \cdot 3n^2x^2 =$

$2r^6m^5 \cdot 10r^2m =$

$3yk \cdot 10y^2k^5 =$

$3p^2v^5 \cdot 4p^5v^5 =$

$3d^4p^2 \cdot 2d^2p =$

$10a^6m^3 \cdot 5a^2m^6 =$

$2f^4n^2 \cdot 4f^5n^5 =$

$3m^3h \cdot 3m^3h^3 =$

$5k^5d^5 \cdot 10kd^6 =$

$4x^5m^2 \cdot 3x^4m^5 =$

$2w^2d^4 \cdot 2w^5d^2 =$

$4w^2b^2 \cdot 4w^6b^2 =$

$5x^3f^3 \cdot 4xf =$

$2f^4q^5 \cdot 4f^6q^2 =$

$5f^3y^2 \cdot 3f^5y =$

$4f^5d^3 \cdot 4f^3d^3 =$

$3g^6h^5 \cdot 3g^5h^6 =$

$5q^5b^3 \cdot 5qb^4 =$

# Answer Key

Simplify each expression.

$2bf^3 \cdot 2b^4f = 4b^5f^4$

$3k^5m^3 \cdot 2k^3m^5 = 6k^8m^8$

$4nx^3 \cdot 3n^2x^2 = 12n^3x^5$

$2r^6m^5 \cdot 10r^2m = 20r^8m^6$

$3yk \cdot 10y^2k^5 = 30y^3k^6$

$3p^2v^5 \cdot 4p^5v^5 = 12p^7v^{10}$

$3d^4p^2 \cdot 2d^2p = 6d^6p^3$

$10a^6m^3 \cdot 5a^2m^6 = 50a^8m^9$

$2f^4n^2 \cdot 4f^5n^5 = 8f^9n^7$

$3m^3h \cdot 3m^3h^3 = 9m^6h^4$

$5k^5d^5 \cdot 10kd^6 = 50k^6d^{11}$

$4x^5m^2 \cdot 3x^4m^5 = 12x^9m^7$

$2w^2d^4 \cdot 2w^5d^2 = 4w^7d^6$

$4w^2b^2 \cdot 4w^6b^2 = 16w^8b^4$

$5x^3f^3 \cdot 4xf = 20x^4f^4$

$2f^4q^5 \cdot 4f^6q^2 = 8f^{10}q^7$

$5f^3y^2 \cdot 3f^5y = 15f^8y^3$

$4f^5d^3 \cdot 4f^3d^3 = 16f^8d^6$

$3g^6h^5 \cdot 3g^5h^6 = 9g^{11}h^{11}$

$5q^5b^3 \cdot 5qb^4 = 25q^6b^7$

Simplify each expression.

$10h^3w^2 \cdot 2h^3w^5 =$

$4a^3u \cdot 4a^2u^3 =$

$4h^6u^6 \cdot 2h^2u^4 =$

$5a^2u^3 \cdot 5a^5u =$

$4xy^3 \cdot 2x^2y^3 =$

$10xu^3 \cdot 2x^5u =$

$3yq \cdot 10y^5q^4 =$

$4tr^6 \cdot 3t^5r^6 =$

$3g^2y^6 \cdot 5g^3y^6 =$

$5h^2w^4 \cdot 2h^2w^4 =$

$4u^6m^4 \cdot 3um^6 =$

$4m^4w \cdot 5m^4w^4 =$

$3n^2q^3 \cdot 4n^6q^4 =$

$5pb^4 \cdot 3p^3b =$

$4r^3g^2 \cdot 4r^3g =$

$10g^3t^6 \cdot 2g^2t^2 =$

$5u^4y^4 \cdot 3uy^2 =$

$5f^3h^6 \cdot 10f^6h =$

$3d^3u^2 \cdot 5d^2u =$

$5y^3t^2 \cdot 3y^4t =$

# Answer Key

Simplify each expression.

$10h^3w^2 \cdot 2h^3w^5 = 20h^6w^7$

$4a^3u \cdot 4a^2u^3 = 16a^5u^4$

$4h^6u^6 \cdot 2h^2u^4 = 8h^8u^{10}$

$5a^2u^3 \cdot 5a^5u = 25a^7u^4$

$4xy^3 \cdot 2x^2y^3 = 8x^3y^6$

$10xu^3 \cdot 2x^5u = 20x^6u^4$

$3yq \cdot 10y^5q^4 = 30y^6q^5$

$4tr^6 \cdot 3t^5r^6 = 12t^6r^{12}$

$3g^2y^6 \cdot 5g^3y^6 = 15g^5y^{12}$

$5h^2w^4 \cdot 2h^2w^4 = 10h^4w^8$

$4u^6m^4 \cdot 3um^6 = 12u^7m^{10}$

$4m^4w \cdot 5m^4w^4 = 20m^8w^5$

$3n^2q^3 \cdot 4n^6q^4 = 12n^8q^7$

$5pb^4 \cdot 3p^3b = 15p^4b^5$

$4r^3g^2 \cdot 4r^3g = 16r^6g^3$

$10g^3t^6 \cdot 2g^2t^2 = 20g^5t^8$

$5u^4y^4 \cdot 3uy^2 = 15u^5y^6$

$5f^3h^6 \cdot 10f^6h = 50f^9h^7$

$3d^3u^2 \cdot 5d^2u = 15d^5u^3$

$5y^3t^2 \cdot 3y^4t = 15y^7t^3$

Simplify each expression.

$10x^2d^4 \cdot 3x^5d^2 =$

$5a^3b^5 \cdot 2ab^4 =$

$10a^4h^2 \cdot 3a^2h^2 =$

$2v^4u^2 \cdot 2v^3u =$

$5v^6h^4 \cdot 10v^4h^3 =$

$3a^2v^2 \cdot 2a^2v =$

$4v^3u^4 \cdot 2v^3u^4 =$

$4k^3g^3 \cdot 4k^2g^6 =$

$3a^3m^4 \cdot 2a^2m^6 =$

$10r^2k^5 \cdot 2r^3k^2 =$

$2u^4k^3 \cdot 10u^6k =$

$3h^4b^4 \cdot 10h^5b^6 =$

$10y^2q^6 \cdot 4y^3q^2 =$

$4vd^2 \cdot 3v^6d^2 =$

$3u^3w \cdot 3u^6w^4 =$

$5ad^2 \cdot 4a^6d^4 =$

$5m^2t \cdot 3m^5t =$

$5u^3w^2 \cdot 3u^3w^2 =$

$2v^2q^3 \cdot 4vq^2 =$

$10p^6r^4 \cdot 3p^3r^3 =$

# Answer Key

Simplify each expression.

$10x^2d^4 \cdot 3x^5d^2 = 30x^7d^6$

$5a^3b^5 \cdot 2ab^4 = 10a^4b^9$

$10a^4h^2 \cdot 3a^2h^2 = 30a^6h^4$

$2v^4u^2 \cdot 2v^3u = 4v^7u^3$

$5v^6h^4 \cdot 10v^4h^3 = 50v^{10}h^7$

$3a^2v^2 \cdot 2a^2v = 6a^4v^3$

$4v^3u^4 \cdot 2v^3u^4 = 8v^6u^8$

$4k^3g^3 \cdot 4k^2g^6 = 16k^5g^9$

$3a^3m^4 \cdot 2a^2m^6 = 6a^5m^{10}$

$10r^2k^5 \cdot 2r^3k^2 = 20r^5k^7$

$2u^4k^3 \cdot 10u^6k = 20u^{10}k^4$

$3h^4b^4 \cdot 10h^5b^6 = 30h^9b^{10}$

$10y^2q^6 \cdot 4y^3q^2 = 40y^5q^8$

$4vd^2 \cdot 3v^6d^2 = 12v^7d^4$

$3u^3w \cdot 3u^6w^4 = 9u^9w^5$

$5ad^2 \cdot 4a^6d^4 = 20a^7d^6$

$5m^2t \cdot 3m^5t = 15m^7t^2$

$5u^3w^2 \cdot 3u^3w^2 = 15u^6w^4$

$2v^2q^3 \cdot 4vq^2 = 8v^3q^5$

$10p^6r^4 \cdot 3p^3r^3 = 30p^9r^7$

Simplify each expression.

$10xr^3 \cdot 3x^5r^2 =$

$4a^5r^2 \cdot 10a^6r^6 =$

$10rq^3 \cdot 3r^3q^5 =$

$3n^4d^5 \cdot 2n^4d =$

$4ht^3 \cdot 2h^2t^2 =$

$2k^4w^2 \cdot 10k^6w^3 =$

$3a^4y^4 \cdot 3a^2y^6 =$

$4x^3a^4 \cdot 3x^3a^2 =$

$10an^6 \cdot 10a^5n^3 =$

$3t^2a^2 \cdot 10t^4a^2 =$

$10ua^4 \cdot 2u^5a^2 =$

$3m^2h^5 \cdot 4m^2h^4 =$

$3n^3q^4 \cdot 4n^3q^2 =$

$10x^2w^2 \cdot 10x^6w =$

$4n^2x^6 \cdot 5n^3x^2 =$

$4v^5w \cdot 3v^4w =$

$10u^5x^6 \cdot 10u^2x^2 =$

$2k^2p \cdot 2k^5p^5 =$

$3p^4k^5 \cdot 10p^5k^6 =$

$3b^6g^6 \cdot 5bg^5 =$

# Answer Key

Simplify each expression.

$10xr^3 \cdot 3x^5r^2 = 30x^6r^5$

$4a^5r^2 \cdot 10a^6r^6 = 40a^{11}r^8$

$10rq^3 \cdot 3r^3q^5 = 30r^4q^8$

$3n^4d^5 \cdot 2n^4d = 6n^8d^6$

$4ht^3 \cdot 2h^2t^2 = 8h^3t^5$

$2k^4w^2 \cdot 10k^6w^3 = 20k^{10}w^5$

$3a^4y^4 \cdot 3a^2y^6 = 9a^6y^{10}$

$4x^3a^4 \cdot 3x^3a^2 = 12x^6a^6$

$10an^6 \cdot 10a^5n^3 = 100a^6n^9$

$3t^2a^2 \cdot 10t^4a^2 = 30t^6a^4$

$10ua^4 \cdot 2u^5a^2 = 20u^6a^6$

$3m^2h^5 \cdot 4m^2h^4 = 12m^4h^9$

$3n^3q^4 \cdot 4n^3q^2 = 12n^6q^6$

$10x^2w^2 \cdot 10x^6w = 100x^8w^3$

$4n^2x^6 \cdot 5n^3x^2 = 20n^5x^8$

$4v^5w \cdot 3v^4w = 12v^9w^2$

$10u^5x^6 \cdot 10u^2x^2 = 100u^7x^8$

$2k^2p \cdot 2k^5p^5 = 4k^7p^6$

$3p^4k^5 \cdot 10p^5k^6 = 30p^9k^{11}$

$3b^6g^6 \cdot 5bg^5 = 15b^7g^{11}$

Simplify each expression.

$3d^5v^4 \cdot 4d^6v^6 =$

$3b^5w^4 \cdot 5b^4w^4 =$

$10x^2a^2 \cdot 3x^2a^6 =$

$3r^3d^3 \cdot 10r^4d^2 =$

$10u^5p^3 \cdot 3up^5 =$

$4u^4q \cdot 5u^4q^5 =$

$4k^4h^2 \cdot 3k^3h^2 =$

$3q^3x^4 \cdot 2qx^2 =$

$3u^6r^3 \cdot 3u^6r^4 =$

$2n^6d^6 \cdot 2n^3d^4 =$

$3b^3d^2 \cdot 3bd^4 =$

$5r^2b^5 \cdot 3r^3b^3 =$

$5k^2p^5 \cdot 2k^3p^3 =$

$2vw \cdot 5v^5w^2 =$

$5r^2t^3 \cdot 4r^6t =$

$10un^4 \cdot 10u^2n^4 =$

$3m^2u^3 \cdot 4m^3u^4 =$

$3u^6n^3 \cdot 4u^5n^3 =$

$5nd^3 \cdot 2n^2d^3 =$

$3u^4a^3 \cdot 5u^2a^2 =$

# Answer Key

Simplify each expression.

$3d^5v^4 \cdot 4d^6v^6 = 12d^{11}v^{10}$

$3b^5w^4 \cdot 5b^4w^4 = 15b^9w^8$

$10x^2a^2 \cdot 3x^2a^6 = 30x^4a^8$

$3r^3d^3 \cdot 10r^4d^2 = 30r^7d^5$

$10u^5p^3 \cdot 3up^5 = 30u^6p^8$

$4u^4q \cdot 5u^4q^5 = 20u^8q^6$

$4k^4h^2 \cdot 3k^3h^2 = 12k^7h^4$

$3q^3x^4 \cdot 2qx^2 = 6q^4x^6$

$3u^6r^3 \cdot 3u^6r^4 = 9u^{12}r^7$

$2n^6d^6 \cdot 2n^3d^4 = 4n^9d^{10}$

$3b^3d^2 \cdot 3bd^4 = 9b^4d^6$

$5r^2b^5 \cdot 3r^3b^3 = 15r^5b^8$

$5k^2p^5 \cdot 2k^3p^3 = 10k^5p^8$

$2vw \cdot 5v^5w^2 = 10v^6w^3$

$5r^2t^3 \cdot 4r^6t = 20r^8t^4$

$10un^4 \cdot 10u^2n^4 = 100u^3n^8$

$3m^2u^3 \cdot 4m^3u^4 = 12m^5u^7$

$3u^6n^3 \cdot 4u^5n^3 = 12u^{11}n^6$

$5nd^3 \cdot 2n^2d^3 = 10n^3d^6$

$3u^4a^3 \cdot 5u^2a^2 = 15u^6a^5$

Simplify each expression.

$4q^4k^4 \cdot 3q^5k =$

$5h^3g^2 \cdot 5h^3g^2 =$

$2x^3t^5 \cdot 2x^5t^3 =$

$2r^2t \cdot 2r^2t =$

$3hp^2 \cdot 3h^2p^2 =$

$2dq^2 \cdot 3d^3q^3 =$

$4q^4p^5 \cdot 5q^3p^3 =$

$5v^4g^3 \cdot 10v^6g =$

$5n^5r^6 \cdot 2n^2r^5 =$

$5ku^5 \cdot 3k^3u^5 =$

$4ry^2 \cdot 5r^6y =$

$2x^6f^3 \cdot 3x^2f^2 =$

$5u^2w^6 \cdot 2uw =$

$5u^6m^5 \cdot 2u^4m^3 =$

$4vu^5 \cdot 5v^6u =$

$10q^3v^4 \cdot 10qv^3 =$

$5m^3a^5 \cdot 4ma^3 =$

$2p^6k^6 \cdot 4p^6k^6 =$

$2a^4h^5 \cdot 2a^2h^6 =$

$10q^3x^5 \cdot 4q^5x^4 =$

# Answer Key

Simplify each expression.

$4q^4k^4 \cdot 3q^5k = 12q^9k^5$

$5h^3g^2 \cdot 5h^3g^2 = 25h^6g^4$

$2x^3t^5 \cdot 2x^5t^3 = 4x^8t^8$

$2r^2t \cdot 2r^2t = 4r^4t^2$

$3hp^2 \cdot 3h^2p^2 = 9h^3p^4$

$2dq^2 \cdot 3d^3q^3 = 6d^4q^5$

$4q^4p^5 \cdot 5q^3p^3 = 20q^7p^8$

$5v^4g^3 \cdot 10v^6g = 50v^{10}g^4$

$5n^5r^6 \cdot 2n^2r^5 = 10n^7r^{11}$

$5ku^5 \cdot 3k^3u^5 = 15k^4u^{10}$

$4ry^2 \cdot 5r^6y = 20r^7y^3$

$2x^6f^3 \cdot 3x^2f^2 = 6x^8f^5$

$5u^2w^6 \cdot 2uw = 10u^3w^7$

$5u^6m^5 \cdot 2u^4m^3 = 10u^{10}m^8$

$4vu^5 \cdot 5v^6u = 20v^7u^6$

$10q^3v^4 \cdot 10qv^3 = 100q^4v^7$

$5m^3a^5 \cdot 4ma^3 = 20m^4a^8$

$2p^6k^6 \cdot 4p^6k^6 = 8p^{12}k^{12}$

$2a^4h^5 \cdot 2a^2h^6 = 4a^6h^{11}$

$10q^3x^5 \cdot 4q^5x^4 = 40q^8x^9$

Simplify each expression.

$3v^3r \cdot 4v^4r =$

$10p^3g^2 \cdot 2p^5g^5 =$

$2q^2d^2 \cdot 10q^3d^5 =$

$5x^3f^3 \cdot 2xf^4 =$

$10r^3h \cdot 4r^2h^5 =$

$4k^3w^3 \cdot 3k^2w^3 =$

$2t^3x^6 \cdot 2t^3x^4 =$

$5t^3u^2 \cdot 3t^3u^3 =$

$3p^5q^6 \cdot 5p^6q^5 =$

$2f^4n^4 \cdot 4f^2n^5 =$

$5k^3a^2 \cdot 4k^4a^6 =$

$10hd \cdot 10h^5d^2 =$

$3ud^2 \cdot 5ud =$

$10a^2p \cdot 3a^2p^2 =$

$3w^6h^6 \cdot 4wh^3 =$

$2f^3r \cdot 3f^6r =$

$3r^2f^5 \cdot 10r^5f^6 =$

$2p^6m^3 \cdot 2p^3m^6 =$

$2k^2a^5 \cdot 5k^4a^3 =$

$2h^3a^2 \cdot 3h^4a^4 =$

# Answer Key

Simplify each expression.

$3v^3r \cdot 4v^4r = 12v^7r^2$

$10p^3g^2 \cdot 2p^5g^5 = 20p^8g^7$

$2q^2d^2 \cdot 10q^3d^5 = 20q^5d^7$

$5x^3f^3 \cdot 2xf^4 = 10x^4f^7$

$10r^3h \cdot 4r^2h^5 = 40r^5h^6$

$4k^3w^3 \cdot 3k^2w^3 = 12k^5w^6$

$2t^3x^6 \cdot 2t^3x^4 = 4t^6x^{10}$

$5t^3u^2 \cdot 3t^3u^3 = 15t^6u^5$

$3p^5q^6 \cdot 5p^6q^5 = 15p^{11}q^{11}$

$2f^4n^4 \cdot 4f^2n^5 = 8f^6n^9$

$5k^3a^2 \cdot 4k^4a^6 = 20k^7a^8$

$10hd \cdot 10h^5d^2 = 100h^6d^3$

$3ud^2 \cdot 5ud = 15u^2d^3$

$10a^2p \cdot 3a^2p^2 = 30a^4p^3$

$3w^6h^6 \cdot 4wh^3 = 12w^7h^9$

$2f^3r \cdot 3f^6r = 6f^9r^2$

$3r^2f^5 \cdot 10r^5f^6 = 30r^7f^{11}$

$2p^6m^3 \cdot 2p^3m^6 = 4p^9m^9$

$2k^2a^5 \cdot 5k^4a^3 = 10k^6a^8$

$2h^3a^2 \cdot 3h^4a^4 = 6h^7a^6$

Simplify each expression.

$3av^4 \cdot 3av^4 =$

$2u^3a^2 \cdot 5u^6a^3 =$

$5g^6u^3 \cdot 5g^5u^5 =$

$3t^5h^6 \cdot 5t^5h^3 =$

$10r^3a^6 \cdot 2r^2a^6 =$

$3k^2v^4 \cdot 10k^3v =$

$2t^6m \cdot 3t^3m^5 =$

$4h^3f^6 \cdot 3h^6f =$

$10p^2w \cdot 5p^5w^5 =$

$10k^3n^5 \cdot 4k^3n =$

$5mb^4 \cdot 3m^6b =$

$3n^3g \cdot 10n^3g^2 =$

$4q^4v^3 \cdot 2q^3v^4 =$

$3h^4r^3 \cdot 2h^6r^3 =$

$3dy^5 \cdot 4d^2y^5 =$

$10h^4t^2 \cdot 5h^3t^5 =$

$3q^4t^2 \cdot 5q^4t^5 =$

$3y^2x \cdot 10y^3x^3 =$

$4b^3r \cdot 4b^3r^5 =$

$2x^6a^3 \cdot 3x^6a =$

# Answer Key

Simplify each expression.

$3av^{4} \cdot 3av^{4} = 9a^{2}v^{8}$

$5g^{6}u^{3} \cdot 5g^{5}u^{5} = 25g^{11}u^{8}$

$10r^{3}a^{6} \cdot 2r^{2}a^{6} = 20r^{5}a^{12}$

$2t^{6}m \cdot 3t^{3}m^{5} = 6t^{9}m^{6}$

$10p^{2}w \cdot 5p^{5}w^{5} = 50p^{7}w^{6}$

$5mb^{4} \cdot 3m^{6}b = 15m^{7}b^{5}$

$4q^{4}v^{3} \cdot 2q^{3}v^{4} = 8q^{7}v^{7}$

$3dy^{5} \cdot 4d^{2}y^{5} = 12d^{3}y^{10}$

$3q^{4}t^{2} \cdot 5q^{4}t^{5} = 15q^{8}t^{7}$

$4b^{3}r \cdot 4b^{3}r^{5} = 16b^{6}r^{6}$

$2u^{3}a^{2} \cdot 5u^{6}a^{3} = 10u^{9}a^{5}$

$3t^{5}h^{6} \cdot 5t^{5}h^{3} = 15t^{10}h^{9}$

$3k^{2}v^{4} \cdot 10k^{3}v = 30k^{5}v^{5}$

$4h^{3}f^{6} \cdot 3h^{6}f = 12h^{9}f^{7}$

$10k^{3}n^{5} \cdot 4k^{3}n = 40k^{6}n^{6}$

$3n^{3}g \cdot 10n^{3}g^{2} = 30n^{6}g^{3}$

$3h^{4}r^{3} \cdot 2h^{6}r^{3} = 6h^{10}r^{6}$

$10h^{4}t^{2} \cdot 5h^{3}t^{5} = 50h^{7}t^{7}$

$3y^{2}x \cdot 10y^{3}x^{3} = 30y^{5}x^{4}$

$2x^{6}a^{3} \cdot 3x^{6}a = 6x^{12}a^{4}$

Simplify each expression.

$10x^3p^4 \cdot 4x^2p^2 =$

$10n^4d^3 \cdot 3nd^4 =$

$2n^5t^6 \cdot 5n^4t^3 =$

$2d^3k^6 \cdot 4d^2k^4 =$

$5vf^3 \cdot 3v^6f^2 =$

$10r^2u^6 \cdot 4r^2u^2 =$

$5b^5a^5 \cdot 10ba^2 =$

$3u^3p^2 \cdot 4u^4p^3 =$

$10h^3d^5 \cdot 5h^2d^5 =$

$5r^4w^2 \cdot 4r^2w =$

$2d^3n^2 \cdot 5d^2n =$

$4m^4v^2 \cdot 3m^5v^6 =$

$2wf^2 \cdot 10w^3f^6 =$

$3v^4q^3 \cdot 2v^4q^5 =$

$3g^3u^4 \cdot 3gu^2 =$

$5g^2h^2 \cdot 3g^3h^2 =$

$2r^3g^5 \cdot 5rg^3 =$

$3d^2g^2 \cdot 3dg^3 =$

$10y^2p^3 \cdot 3y^2p^2 =$

$3r^2g^3 \cdot 4r^3g^3 =$

# Answer Key

Simplify each expression.

$10x^3p^4 \cdot 4x^2p^2 = 40x^5p^6$

$10n^4d^3 \cdot 3nd^4 = 30n^5d^7$

$2n^5t^6 \cdot 5n^4t^3 = 10n^9t^9$

$2d^3k^6 \cdot 4d^2k^4 = 8d^5k^{10}$

$5vf^3 \cdot 3v^6f^2 = 15v^7f^5$

$10r^2u^6 \cdot 4r^2u^2 = 40r^4u^8$

$5b^5a^5 \cdot 10ba^2 = 50b^6a^7$

$3u^3p^2 \cdot 4u^4p^3 = 12u^7p^5$

$10h^3d^5 \cdot 5h^2d^5 = 50h^5d^{10}$

$5r^4w^2 \cdot 4r^2w = 20r^6w^3$

$2d^3n^2 \cdot 5d^2n = 10d^5n^3$

$4m^4v^2 \cdot 3m^5v^6 = 12m^9v^8$

$2wf^2 \cdot 10w^3f^6 = 20w^4f^8$

$3v^4q^3 \cdot 2v^4q^5 = 6v^8q^8$

$3g^3u^4 \cdot 3gu^2 = 9g^4u^6$

$5g^2h^2 \cdot 3g^3h^2 = 15g^5h^4$

$2r^3g^5 \cdot 5rg^3 = 10r^4g^8$

$3d^2g^2 \cdot 3dg^3 = 9d^3g^5$

$10y^2p^3 \cdot 3y^2p^2 = 30y^4p^5$

$3r^2g^3 \cdot 4r^3g^3 = 12r^5g^6$

Simplify each expression.

$2y^6q^2 \cdot 3y^5q^6 =$

$10a^2y \cdot 3a^3y^2 =$

$3m^3p^3 \cdot 5m^5p^3 =$

$2r^4a^5 \cdot 4ra^2 =$

$4vb^2 \cdot 2v^5b =$

$4g^6y \cdot 2g^5y^6 =$

$10h^4v^3 \cdot 2h^3v^3 =$

$10g^5k^4 \cdot 2g^3k^6 =$

$10n^5d^3 \cdot 2n^4d^4 =$

$3bk^6 \cdot 3b^3k^5 =$

$10q^3h \cdot 10q^6h^5 =$

$3g^2r \cdot 3gr^3 =$

$4m^4v^4 \cdot 4m^3v^2 =$

$10g^2m^5 \cdot 2g^5m =$

$2pq \cdot 4p^6q =$

$10vt^3 \cdot 4v^3t^2 =$

$2g^3v^5 \cdot 2g^5v^4 =$

$2f^5w^2 \cdot 2f^3w^4 =$

$4q^2g^5 \cdot 4qg^3 =$

$2t^3u^5 \cdot 4t^5u^2 =$

# Answer Key

Simplify each expression.

$2y^6q^2 \cdot 3y^5q^6 = 6y^{11}q^8$

$10a^2y \cdot 3a^3y^2 = 30a^5y^3$

$3m^3p^3 \cdot 5m^5p^3 = 15m^8p^6$

$2r^4a^5 \cdot 4ra^2 = 8r^5a^7$

$4vb^2 \cdot 2v^5b = 8v^6b^3$

$4g^6y \cdot 2g^5y^6 = 8g^{11}y^7$

$10h^4v^3 \cdot 2h^3v^3 = 20h^7v^6$

$10g^5k^4 \cdot 2g^3k^6 = 20g^8k^{10}$

$10n^5d^3 \cdot 2n^4d^4 = 20n^9d^7$

$3bk^6 \cdot 3b^3k^5 = 9b^4k^{11}$

$10q^3h \cdot 10q^6h^5 = 100q^9h^6$

$3g^2r \cdot 3gr^3 = 9g^3r^4$

$4m^4v^4 \cdot 4m^3v^2 = 16m^7v^6$

$10g^2m^5 \cdot 2g^5m = 20g^7m^6$

$2pq \cdot 4p^6q = 8p^7q^2$

$10vt^3 \cdot 4v^3t^2 = 40v^4t^5$

$2g^3v^5 \cdot 2g^5v^4 = 4g^8v^9$

$2f^5w^2 \cdot 2f^3w^4 = 4f^8w^6$

$4q^2g^5 \cdot 4qg^3 = 16q^3g^8$

$2t^3u^5 \cdot 4t^5u^2 = 8t^8u^7$

Simplify each expression.

$10g^{4}r^{5} \cdot 3g^{3}r^{4} =$

$2y^{2}x^{5} \cdot 4yx^{3} =$

$3qh^{5} \cdot 3q^{2}h^{2} =$

$4g^{3}p \cdot 2g^{2}p =$

$2k^{6}x^{4} \cdot 5k^{6}x^{2} =$

$2rm^{5} \cdot 5r^{6}m^{4} =$

$4p^{5}q^{6} \cdot 3p^{2}q^{6} =$

$4q^{6}k^{4} \cdot 5q^{5}k^{2} =$

$3u^{2}y^{2} \cdot 5u^{2}y^{5} =$

$4m^{6}t^{4} \cdot 3m^{2}t =$

$5p^{4}w^{3} \cdot 3p^{3}w^{3} =$

$2y^{3}d \cdot 3y^{3}d^{4} =$

$5x^{4}p^{2} \cdot 2x^{4}p^{4} =$

$3d^{6}x^{6} \cdot 3d^{4}x^{4} =$

$2k^{5}d^{3} \cdot 2k^{5}d^{5} =$

$4h^{2}d^{6} \cdot 4h^{6}d =$

$10u^{4}g^{6} \cdot 4u^{2}g^{6} =$

$5y^{6}w^{6} \cdot 10y^{3}w^{4} =$

$3t^{5}r \cdot 2t^{6}r^{4} =$

$10n^{2}t^{4} \cdot 2n^{5}t =$

# Answer Key

Simplify each expression.

$10g^4r^5 \cdot 3g^3r^4 = 30g^7r^9$

$2y^2x^5 \cdot 4yx^3 = 8y^3x^8$

$3qh^5 \cdot 3q^2h^2 = 9q^3h^7$

$4g^3p \cdot 2g^2p = 8g^5p^2$

$2k^6x^4 \cdot 5k^6x^2 = 10k^{12}x^6$

$2rm^5 \cdot 5r^6m^4 = 10r^7m^9$

$4p^5q^6 \cdot 3p^2q^6 = 12p^7q^{12}$

$4q^6k^4 \cdot 5q^5k^2 = 20q^{11}k^6$

$3u^2y^2 \cdot 5u^2y^5 = 15u^4y^7$

$4m^6t^4 \cdot 3m^2t = 12m^8t^5$

$5p^4w^3 \cdot 3p^3w^3 = 15p^7w^6$

$2y^3d \cdot 3y^3d^4 = 6y^6d^5$

$5x^4p^2 \cdot 2x^4p^4 = 10x^8p^6$

$3d^6x^6 \cdot 3d^4x^4 = 9d^{10}x^{10}$

$2k^5d^3 \cdot 2k^5d^5 = 4k^{10}d^8$

$4h^2d^6 \cdot 4h^6d = 16h^8d^7$

$10u^4g^6 \cdot 4u^2g^6 = 40u^6g^{12}$

$5y^6w^6 \cdot 10y^3w^4 = 50y^9w^{10}$

$3t^5r \cdot 2t^6r^4 = 6t^{11}r^5$

$10n^2t^4 \cdot 2n^5t = 20n^7t^5$

Simplify each expression.

$2x^3q^2 \cdot 2x^4q^5 =$

$10b^4y^5 \cdot 3b^4y^2 =$

$5p^3h \cdot 4p^3h^3 =$

$10a^5b^2 \cdot 10a^3b^6 =$

$3w^3m^3 \cdot 3w^4m =$

$3g^4v^6 \cdot 10g^6v^6 =$

$4t^3u^4 \cdot 2tu^3 =$

$5uq \cdot 3u^2q^3 =$

$3k^3q^6 \cdot 5k^6q^2 =$

$3k^3u \cdot 10k^3u^3 =$

$5q^3x^2 \cdot 3qx^2 =$

$10g^3h^2 \cdot 4g^2h =$

$2v^3t^5 \cdot 5v^3t^5 =$

$4q^3n \cdot 4q^5n^5 =$

$4a^6y^3 \cdot 5ay =$

$2u^4n \cdot 3u^3n^2 =$

$10y^5r^4 \cdot 3y^3r^6 =$

$3ta^5 \cdot 3t^2a^3 =$

$3k^6d \cdot 4k^2d^2 =$

$4a^6b^3 \cdot 3a^2b^5 =$

# Answer Key

Simplify each expression.

$2x^3q^2 \cdot 2x^4q^5 = 4x^7q^7$

$10b^4y^5 \cdot 3b^4y^2 = 30b^8y^7$

$5p^3h \cdot 4p^3h^3 = 20p^6h^4$

$10a^5b^2 \cdot 10a^3b^6 = 100a^8b^8$

$3w^3m^3 \cdot 3w^4m = 9w^7m^4$

$3g^4v^6 \cdot 10g^6v^6 = 30g^{10}v^{12}$

$4t^3u^4 \cdot 2tu^3 = 8t^4u^7$

$5uq \cdot 3u^2q^3 = 15u^3q^4$

$3k^3q^6 \cdot 5k^6q^2 = 15k^9q^8$

$3k^3u \cdot 10k^3u^3 = 30k^6u^4$

$5q^3x^2 \cdot 3qx^2 = 15q^4x^4$

$10g^3h^2 \cdot 4g^2h = 40g^5h^3$

$2v^3t^5 \cdot 5v^3t^5 = 10v^6t^{10}$

$4q^3n \cdot 4q^5n^5 = 16q^8n^6$

$4a^6y^3 \cdot 5ay = 20a^7y^4$

$2u^4n \cdot 3u^3n^2 = 6u^7n^3$

$10y^5r^4 \cdot 3y^3r^6 = 30y^8r^{10}$

$3ta^5 \cdot 3t^2a^3 = 9t^3a^8$

$3k^6d \cdot 4k^2d^2 = 12k^8d^3$

$4a^6b^3 \cdot 3a^2b^5 = 12a^8b^8$

Simplify each expression.

$10p^5m^3 \cdot 4p^2m^2 =$

$10t^3w^3 \cdot 2tw^4 =$

$2p^6t^2 \cdot 5p^3t =$

$2yw^4 \cdot 2y^6w^3 =$

$10a^5g^4 \cdot 4a^5g =$

$4f^3v \cdot 2f^5v^2 =$

$5bt^4 \cdot 3b^3t^3 =$

$2rq^3 \cdot 3rq^6 =$

$5v^3y^3 \cdot 3v^2y^3 =$

$5f^5k^6 \cdot 5fk^6 =$

$5n^3w \cdot 2n^4w =$

$3y^5f \cdot 3y^4f^4 =$

$10mv^4 \cdot 2m^5v^2 =$

$5t^3f^6 \cdot 3t^6f^6 =$

$10p^6u^2 \cdot 5p^2u^2 =$

$3n^6g^4 \cdot 10n^3g^3 =$

$4d^5u^5 \cdot 4d^2u =$

$4v^4u^5 \cdot 4v^6u^5 =$

$3y^2g \cdot 10yg^3 =$

$3u^2g^2 \cdot 4ug^4 =$

# Answer Key

Simplify each expression.

$10p^5m^3 \cdot 4p^2m^2 = 40p^7m^5$

$10t^3w^3 \cdot 2tw^4 = 20t^4w^7$

$2p^6t^2 \cdot 5p^3t = 10p^9t^3$

$2yw^4 \cdot 2y^6w^3 = 4y^7w^7$

$10a^5g^4 \cdot 4a^5g = 40a^{10}g^5$

$4f^3v \cdot 2f^5v^2 = 8f^8v^3$

$5bt^4 \cdot 3b^3t^3 = 15b^4t^7$

$2rq^3 \cdot 3rq^6 = 6r^2q^9$

$5v^3y^3 \cdot 3v^2y^3 = 15v^5y^6$

$5f^5k^6 \cdot 5fk^6 = 25f^6k^{12}$

$5n^3w \cdot 2n^4w = 10n^7w^2$

$3y^5f \cdot 3y^4f^4 = 9y^9f^5$

$10mv^4 \cdot 2m^5v^2 = 20m^6v^6$

$5t^3f^6 \cdot 3t^6f^6 = 15t^9f^{12}$

$10p^6u^2 \cdot 5p^2u^2 = 50p^8u^4$

$3n^6g^4 \cdot 10n^3g^3 = 30n^9g^7$

$4d^5u^5 \cdot 4d^2u = 16d^7u^6$

$4v^4u^5 \cdot 4v^6u^5 = 16v^{10}u^{10}$

$3y^2g \cdot 10yg^3 = 30y^3g^4$

$3u^2g^2 \cdot 4ug^4 = 12u^3g^6$

Simplify each expression.

$\frac{6f^3}{4f^5} =$

$\frac{40p^4}{50p^2} =$

$\frac{9k^3}{6k^2} =$

$\frac{6b^4}{8b} =$

$\frac{15r^3}{3r^5} =$

$\frac{20h}{8h^6} =$

$\frac{20r^3}{50r^5} =$

$\frac{8f}{6f^6} =$

$\frac{50v^3}{40v^6} =$

$\frac{4w^2}{6w^2} =$

$\frac{6f^3}{15f} =$

$\frac{6f}{2f^6} =$

$\frac{40x^5}{30x} =$

$\frac{12p}{20p^3} =$

$\frac{12p^6}{4p^2} =$

$\frac{9q}{12q^5} =$

$\frac{16y^4}{12y} =$

$\frac{4w^5}{12w^6} =$

$\frac{5f^5}{15f^3} =$

$\frac{6m^3}{10m^5} =$

# Answer Key

Simplify each expression.

$\frac{6f^3}{4f^5} = \frac{3}{2f^2}$

$\frac{40p^4}{50p^2} = \frac{4p^2}{5}$

$\frac{9k^3}{6k^2} = \frac{3k}{2}$

$\frac{6b^4}{8b} = \frac{3b^3}{4}$

$\frac{15r^3}{3r^5} = \frac{5}{r^2}$

$\frac{20h}{8h^6} = \frac{5}{2h^5}$

$\frac{20r^3}{50r^5} = \frac{2}{5r^2}$

$\frac{8f}{6f^6} = \frac{4}{3f^5}$

$\frac{50v^3}{40v^6} = \frac{5}{4v^3}$

$\frac{4w^2}{6w^2} = \frac{2}{3}$

$\frac{6f^3}{15f} = \frac{2f^2}{5}$

$\frac{6f}{2f^6} = \frac{3}{f^5}$

$\frac{40x^5}{30x} = \frac{4x^4}{3}$

$\frac{12p}{20p^3} = \frac{3}{5p^2}$

$\frac{12p^6}{4p^2} = 3p^4$

$\frac{9q}{12q^5} = \frac{3}{4q^4}$

$\frac{16y^4}{12y} = \frac{4y^3}{3}$

$\frac{4w^5}{12w^6} = \frac{1}{3w}$

$\frac{5f^5}{15f^3} = \frac{f^2}{3}$

$\frac{6m^3}{10m^5} = \frac{3}{5m^2}$

Simplify each expression.

$\frac{5d^3}{20d^3} =$

$\frac{30k^2}{10k} =$

$\frac{12n^5}{20n} =$

$\frac{50k^3}{20k^5} =$

$\frac{5x^4}{25x} =$

$\frac{5x^3}{20x^2} =$

$\frac{3t^2}{6t^2} =$

$\frac{20x}{4x^4} =$

$\frac{30g^5}{40g} =$

$\frac{20p^2}{16p^6} =$

$\frac{15b^4}{9b^2} =$

$\frac{9b^6}{12b^2} =$

$\frac{20k^4}{25k^6} =$

$\frac{4m^3}{40m} =$

$\frac{10u^3}{5u^3} =$

$\frac{6u^2}{8u^4} =$

$\frac{100w^6}{10w^2} =$

$\frac{30t^2}{10t^2} =$

$\frac{5g^4}{25g^2} =$

$\frac{10f^2}{6f} =$

# Answer Key

Simplify each expression.

$\frac{5d^3}{20d^3} = \frac{1}{4}$

$\frac{15b^4}{9b^2} = \frac{5b^2}{3}$

$\frac{30k^2}{10k} = 3k$

$\frac{9b^6}{12b^2} = \frac{3b^4}{4}$

$\frac{12n^5}{20n} = \frac{3n^4}{5}$

$\frac{20k^4}{25k^6} = \frac{4}{5k^2}$

$\frac{50k^3}{20k^5} = \frac{5}{2k^2}$

$\frac{4m^3}{40m} = \frac{m^2}{10}$

$\frac{5x^4}{25x} = \frac{x^3}{5}$

$\frac{10u^3}{5u^3} = 2$

$\frac{5x^3}{20x^2} = \frac{x}{4}$

$\frac{6u^2}{8u^4} = \frac{3}{4u^2}$

$\frac{3t^2}{6t^2} = \frac{1}{2}$

$\frac{100w^6}{10w^2} = 10w^4$

$\frac{20x}{4x^4} = \frac{5}{x^3}$

$\frac{30t^2}{10t^2} = 3$

$\frac{30g^5}{40g} = \frac{3g^4}{4}$

$\frac{5g^4}{25g^2} = \frac{g^2}{5}$

$\frac{20p^2}{16p^6} = \frac{5}{4p^4}$

$\frac{10f^2}{6f} = \frac{5f}{3}$

Simplify each expression.

$\frac{8t^4}{20t^2} =$

$\frac{6n^4}{9n^4} =$

$\frac{12v}{8v^6} =$

$\frac{5a^6}{20a^2} =$

$\frac{3x^6}{6x^6} =$

$\frac{40g}{30g^5} =$

$\frac{8q}{4q^3} =$

$\frac{40t}{30t^2} =$

$\frac{30x^2}{40x^2} =$

$\frac{12d^5}{16d^5} =$

$\frac{30f^4}{50f^2} =$

$\frac{20u^5}{2u^6} =$

$\frac{2v}{8v^6} =$

$\frac{10n}{25n^4} =$

$\frac{20y^3}{50y^6} =$

$\frac{5x^2}{15x^4} =$

$\frac{20a^4}{16a^2} =$

$\frac{5f^2}{50f^2} =$

$\frac{10t}{15t^2} =$

$\frac{6t^2}{15t^3} =$

# Answer Key

Simplify each expression.

$\frac{8t^4}{20t^2} = \frac{2t^2}{5}$

$\frac{30f^4}{50f^2} = \frac{3f^2}{5}$

$\frac{6n^4}{9n^4} = \frac{2}{3}$

$\frac{20u^5}{2u^6} = \frac{10}{u}$

$\frac{12v}{8v^6} = \frac{3}{2v^5}$

$\frac{2v}{8v^6} = \frac{1}{4v^5}$

$\frac{5a^6}{20a^2} = \frac{a^4}{4}$

$\frac{10n}{25n^4} = \frac{2}{5n^3}$

$\frac{3x^6}{6x^6} = \frac{1}{2}$

$\frac{20y^3}{50y^6} = \frac{2}{5y^3}$

$\frac{40g}{30g^5} = \frac{4}{3g^4}$

$\frac{5x^2}{15x^4} = \frac{1}{3x^2}$

$\frac{8q}{4q^3} = \frac{2}{q^2}$

$\frac{20a^4}{16a^2} = \frac{5a^2}{4}$

$\frac{40t}{30t^2} = \frac{4}{3t}$

$\frac{5f^2}{50f^2} = \frac{1}{10}$

$\frac{30x^2}{40x^2} = \frac{3}{4}$

$\frac{10t}{15t^2} = \frac{2}{3t}$

$\frac{12d^5}{16d^5} = \frac{3}{4}$

$\frac{6t^2}{15t^3} = \frac{2}{5t}$

Simplify each expression.

$\frac{10u^2}{15u^3} =$

$\frac{2y^3}{6y^2} =$

$\frac{16w^2}{4w^5} =$

$\frac{20f^4}{2f^4} =$

$\frac{15x^4}{25x^4} =$

$\frac{15k}{25k^4} =$

$\frac{12d^6}{15d^3} =$

$\frac{50m}{30m^6} =$

$\frac{25f^5}{20f} =$

$\frac{10a}{15a^3} =$

$\frac{12n^3}{8n^3} =$

$\frac{8y^5}{2y^2} =$

$\frac{25m}{10m^5} =$

$\frac{30n^5}{50n^3} =$

$\frac{25y^2}{10y^2} =$

$\frac{4w^3}{2w^5} =$

$\frac{4a}{2a} =$

$\frac{20u}{50u^6} =$

$\frac{12x^2}{4x^2} =$

$\frac{10n^6}{50n^6} =$

# Answer Key

Simplify each expression.

$\frac{10u^2}{15u^3} = \frac{2}{3u}$

$\frac{2y^3}{6y^2} = \frac{y}{3}$

$\frac{16w^2}{4w^5} = \frac{4}{w^3}$

$\frac{20f^4}{2f^4} = 10$

$\frac{15x^4}{25x^4} = \frac{3}{5}$

$\frac{15k}{25k^4} = \frac{3}{5k^3}$

$\frac{12d^6}{15d^3} = \frac{4d^3}{5}$

$\frac{50m}{30m^6} = \frac{5}{3m^5}$

$\frac{25f^5}{20f} = \frac{5f^4}{4}$

$\frac{10a}{15a^3} = \frac{2}{3a^2}$

$\frac{12n^3}{8n^3} = \frac{3}{2}$

$\frac{8y^5}{2y^2} = 4y^3$

$\frac{25m}{10m^5} = \frac{5}{2m^4}$

$\frac{30n^5}{50n^3} = \frac{3n^2}{5}$

$\frac{25y^2}{10y^2} = \frac{5}{2}$

$\frac{4w^3}{2w^5} = \frac{2}{w^2}$

$\frac{4a}{2a} = 2$

$\frac{20u}{50u^6} = \frac{2}{5u^5}$

$\frac{12x^2}{4x^2} = 3$

$\frac{10n^6}{50n^6} = \frac{1}{5}$

Simplify each expression.

$\frac{2d}{6d^2} =$

$\frac{4k}{16k^5} =$

$\frac{9w^6}{15w^6} =$

$\frac{15x^5}{12x^6} =$

$\frac{8r^2}{4r^3} =$

$\frac{5d^4}{20d^5} =$

$\frac{10v}{6v^3} =$

$\frac{15h^3}{3h^6} =$

$\frac{20u}{4u^3} =$

$\frac{6d^6}{4d^5} =$

$\frac{15m^3}{20m^2} =$

$\frac{25d}{5d^3} =$

$\frac{10t}{8t^3} =$

$\frac{30u^4}{3u^5} =$

$\frac{6h^5}{2h^4} =$

$\frac{10r}{20r} =$

$\frac{40p^3}{4p^2} =$

$\frac{15x^2}{20x^3} =$

$\frac{10u^2}{25u^3} =$

$\frac{4f^4}{40f^4} =$

# Answer Key

Simplify each expression.

$\frac{2d}{6d^2} = \frac{1}{3d}$

$\frac{4k}{16k^5} = \frac{1}{4k^4}$

$\frac{9w^6}{15w^6} = \frac{3}{5}$

$\frac{15x^5}{12x^6} = \frac{5}{4x}$

$\frac{8r^2}{4r^3} = \frac{2}{r}$

$\frac{5d^4}{20d^5} = \frac{1}{4d}$

$\frac{10v}{6v^3} = \frac{5}{3v^2}$

$\frac{15h^3}{3h^6} = \frac{5}{h^3}$

$\frac{20u}{4u^3} = \frac{5}{u^2}$

$\frac{6d^6}{4d^5} = \frac{3d}{2}$

$\frac{15m^3}{20m^2} = \frac{3m}{4}$

$\frac{25d}{5d^3} = \frac{5}{d^2}$

$\frac{10t}{8t^3} = \frac{5}{4t^2}$

$\frac{30u^4}{3u^5} = \frac{10}{u}$

$\frac{6h^5}{2h^4} = 3h$

$\frac{10r}{20r} = \frac{1}{2}$

$\frac{40p^3}{4p^2} = 10p$

$\frac{15x^2}{20x^3} = \frac{3}{4x}$

$\frac{10u^2}{25u^3} = \frac{2}{5u}$

$\frac{4f^4}{40f^4} = \frac{1}{10}$

Simplify each expression.

$\frac{9a^2}{6a^6} =$

$\frac{4y^6}{6y^5} =$

$\frac{8w^2}{10w^6} =$

$\frac{2u^2}{6u^2} =$

$\frac{10m^3}{50m^4} =$

$\frac{5b^3}{25b^3} =$

$\frac{9t^3}{12t} =$

$\frac{3w^3}{15w^2} =$

$\frac{4w}{12w^6} =$

$\frac{3u^5}{6u^5} =$

$\frac{20w^3}{8w^5} =$

$\frac{15f^4}{6f^5} =$

$\frac{4h^6}{12h} =$

$\frac{2t^2}{8t^2} =$

$\frac{20g^6}{50g} =$

$\frac{9q^2}{15q^5} =$

$\frac{6f^5}{3f^2} =$

$\frac{20p}{5p^3} =$

$\frac{30u^3}{50u} =$

$\frac{12d^3}{15d^2} =$

# Answer Key

Simplify each expression.

$\frac{9a^2}{6a^6} = \frac{3}{2a^4}$

$\frac{20w^3}{8w^5} = \frac{5}{2w^2}$

$\frac{4y^6}{6y^5} = \frac{2y}{3}$

$\frac{15f^4}{6f^5} = \frac{5}{2f}$

$\frac{8w^2}{10w^6} = \frac{4}{5w^4}$

$\frac{4h^6}{12h} = \frac{h^5}{3}$

$\frac{2u^2}{6u^2} = \frac{1}{3}$

$\frac{2t^2}{8t^2} = \frac{1}{4}$

$\frac{10m^3}{50m^4} = \frac{1}{5m}$

$\frac{20g^6}{50g} = \frac{2g^5}{5}$

$\frac{5b^3}{25b^3} = \frac{1}{5}$

$\frac{9q^2}{15q^5} = \frac{3}{5q^3}$

$\frac{9t^3}{12t} = \frac{3t^2}{4}$

$\frac{6f^5}{3f^2} = 2f^3$

$\frac{3w^3}{15w^2} = \frac{w}{5}$

$\frac{20p}{5p^3} = \frac{4}{p^2}$

$\frac{4w}{12w^6} = \frac{1}{3w^5}$

$\frac{30u^3}{50u} = \frac{3u^2}{5}$

$\frac{3u^5}{6u^5} = \frac{1}{2}$

$\frac{12d^3}{15d^2} = \frac{4d}{5}$

Simplify each expression.

$\frac{4x^3}{6x^4} =$

$\frac{20y^2}{50y^3} =$

$\frac{5w^2}{50w^6} =$

$\frac{50t^2}{30t^5} =$

$\frac{25f^3}{15f^4} =$

$\frac{10y^2}{5y} =$

$\frac{20x^6}{15x^4} =$

$\frac{20w}{10w} =$

$\frac{10w^4}{30w} =$

$\frac{25p^2}{15p^3} =$

$\frac{8f^3}{10f^6} =$

$\frac{4f^3}{10f^3} =$

$\frac{30t^3}{10t} =$

$\frac{9a^2}{15a^4} =$

$\frac{30a^4}{3a^6} =$

$\frac{6t}{9t^6} =$

$\frac{12v^3}{16v} =$

$\frac{15h^2}{5h^6} =$

$\frac{30u^3}{10u^3} =$

$\frac{15k^2}{12k^3} =$

# Answer Key

Simplify each expression.

$\frac{4x^3}{6x^4} = \frac{2}{3x}$

$\frac{5w^2}{50w^6} = \frac{1}{10w^4}$

$\frac{25f^3}{15f^4} = \frac{5}{3f}$

$\frac{20x^6}{15x^4} = \frac{4x^2}{3}$

$\frac{10w^4}{30w} = \frac{w^3}{3}$

$\frac{8f^3}{10f^6} = \frac{4}{5f^3}$

$\frac{30t^3}{10t} = 3t^2$

$\frac{30a^4}{3a^6} = \frac{10}{a^2}$

$\frac{12v^3}{16v} = \frac{3v^2}{4}$

$\frac{30u^3}{10u^3} = 3$

$\frac{20y^2}{50y^3} = \frac{2}{5y}$

$\frac{50t^2}{30t^5} = \frac{5}{3t^3}$

$\frac{10y^2}{5y} = 2y$

$\frac{20w}{10w} = 2$

$\frac{25p^2}{15p^3} = \frac{5}{3p}$

$\frac{4f^3}{10f^3} = \frac{2}{5}$

$\frac{9a^2}{15a^4} = \frac{3}{5a^2}$

$\frac{6t}{9t^6} = \frac{2}{3t^5}$

$\frac{15h^2}{5h^6} = \frac{3}{h^4}$

$\frac{15k^2}{12k^3} = \frac{5}{4k}$

Simplify each expression.

$\frac{10x^3}{6x^2} =$

$\frac{8t^3}{10t^3} =$

$\frac{3x^5}{6x^3} =$

$\frac{10u^3}{50u^3} =$

$\frac{30k^6}{40k} =$

$\frac{15m^5}{6m^2} =$

$\frac{9m^2}{12m^2} =$

$\frac{9d^6}{6d^2} =$

$\frac{15d^5}{9d^3} =$

$\frac{15w^2}{10w^4} =$

$\frac{16b^3}{12b^6} =$

$\frac{15h^4}{10h^3} =$

$\frac{10d^2}{50d} =$

$\frac{16f^5}{12f^5} =$

$\frac{6v^3}{8v^6} =$

$\frac{50n^3}{5n^6} =$

$\frac{9p^3}{15p} =$

$\frac{4f^3}{40f^3} =$

$\frac{20p}{5p^6} =$

$\frac{15k^5}{5k^2} =$

# Answer Key

Simplify each expression.

$$\frac{10x^3}{6x^2} = \frac{5x}{3}$$

$$\frac{3x^5}{6x^3} = \frac{x^2}{2}$$

$$\frac{30k^6}{40k} = \frac{3k^5}{4}$$

$$\frac{9m^2}{12m^2} = \frac{3}{4}$$

$$\frac{15d^5}{9d^3} = \frac{5d^2}{3}$$

$$\frac{16b^3}{12b^6} = \frac{4}{3b^3}$$

$$\frac{10d^2}{50d} = \frac{d}{5}$$

$$\frac{6v^3}{8v^6} = \frac{3}{4v^3}$$

$$\frac{9p^3}{15p} = \frac{3p^2}{5}$$

$$\frac{20p}{5p^6} = \frac{4}{p^5}$$

$$\frac{8t^3}{10t^3} = \frac{4}{5}$$

$$\frac{10u^3}{50u^3} = \frac{1}{5}$$

$$\frac{15m^5}{6m^2} = \frac{5m^3}{2}$$

$$\frac{9d^6}{6d^2} = \frac{3d^4}{2}$$

$$\frac{15w^2}{10w^4} = \frac{3}{2w^2}$$

$$\frac{15h^4}{10h^3} = \frac{3h}{2}$$

$$\frac{16f^5}{12f^5} = \frac{4}{3}$$

$$\frac{50n^3}{5n^6} = \frac{10}{n^3}$$

$$\frac{4f^3}{40f^3} = \frac{1}{10}$$

$$\frac{15k^5}{5k^2} = 3k^3$$

Simplify each expression.

$\frac{10d^2}{15d} =$

$\frac{6r}{9r^2} =$

$\frac{40v^2}{50v^3} =$

$\frac{16h^3}{4h^5} =$

$\frac{15r^5}{20r^2} =$

$\frac{20w^4}{30w^2} =$

$\frac{30a}{20a^4} =$

$\frac{4t^5}{6t^2} =$

$\frac{20k^6}{10k^4} =$

$\frac{15y^2}{10y^3} =$

$\frac{15k^6}{9k^5} =$

$\frac{20k^3}{16k} =$

$\frac{4k^4}{8k^2} =$

$\frac{10f^2}{4f^3} =$

$\frac{25g^2}{15g^5} =$

$\frac{5u}{50u^2} =$

$\frac{8g^5}{12g^5} =$

$\frac{3v^2}{12v^2} =$

$\frac{4q^3}{16q^2} =$

$\frac{10g^6}{30g} =$

# Answer Key

Simplify each expression.

$\frac{10d^2}{15d} = \frac{2d}{3}$

$\frac{6r}{9r^2} = \frac{2}{3r}$

$\frac{40v^2}{50v^3} = \frac{4}{5v}$

$\frac{16h^3}{4h^5} = \frac{4}{h^2}$

$\frac{15r^5}{20r^2} = \frac{3r^3}{4}$

$\frac{20w^4}{30w^2} = \frac{2w^2}{3}$

$\frac{30a}{20a^4} = \frac{3}{2a^3}$

$\frac{4t^5}{6t^2} = \frac{2t^3}{3}$

$\frac{20k^6}{10k^4} = 2k^2$

$\frac{15y^2}{10y^3} = \frac{3}{2y}$

$\frac{15k^6}{9k^5} = \frac{5k}{3}$

$\frac{20k^3}{16k} = \frac{5k^2}{4}$

$\frac{4k^4}{8k^2} = \frac{k^2}{2}$

$\frac{10f^2}{4f^3} = \frac{5}{2f}$

$\frac{25g^2}{15g^5} = \frac{5}{3g^3}$

$\frac{5u}{50u^2} = \frac{1}{10u}$

$\frac{8g^5}{12g^5} = \frac{2}{3}$

$\frac{3v^2}{12v^2} = \frac{1}{4}$

$\frac{4q^3}{16q^2} = \frac{q}{4}$

$\frac{10g^6}{30g} = \frac{g^5}{3}$

Simplify each expression.

$\frac{8y^2}{10y^4} =$

$\frac{5b}{20b} =$

$\frac{3g^6}{30g^6} =$

$\frac{20h^2}{4h^5} =$

$\frac{5m^4}{20m^2} =$

$\frac{5t}{20t^3} =$

$\frac{10v^4}{25v^2} =$

$\frac{50m^6}{5m^2} =$

$\frac{2q^5}{20q^4} =$

$\frac{4t}{20t^3} =$

$\frac{10a^6}{8a^6} =$

$\frac{4m^5}{2m} =$

$\frac{20a}{2a^3} =$

$\frac{6u^2}{9u^6} =$

$\frac{9v}{15v^5} =$

$\frac{16q^2}{20q^3} =$

$\frac{15d^5}{9d} =$

$\frac{20w^5}{50w^5} =$

$\frac{50y^2}{20y^4} =$

$\frac{8u^4}{20u^5} =$

# Answer Key

Simplify each expression.

$$\frac{8y^2}{10y^4} = \frac{4}{5y^2}$$

$$\frac{5b}{20b} = \frac{1}{4}$$

$$\frac{3g^6}{30g^6} = \frac{1}{10}$$

$$\frac{20h^2}{4h^5} = \frac{5}{h^3}$$

$$\frac{5m^4}{20m^2} = \frac{m^2}{4}$$

$$\frac{5t}{20t^3} = \frac{1}{4t^2}$$

$$\frac{10v^4}{25v^2} = \frac{2v^2}{5}$$

$$\frac{50m^6}{5m^2} = 10m^4$$

$$\frac{2q^5}{20q^4} = \frac{q}{10}$$

$$\frac{4t}{20t^3} = \frac{1}{5t^2}$$

$$\frac{10a^6}{8a^6} = \frac{5}{4}$$

$$\frac{4m^5}{2m} = 2m^4$$

$$\frac{20a}{2a^3} = \frac{10}{a^2}$$

$$\frac{6u^2}{9u^6} = \frac{2}{3u^4}$$

$$\frac{9v}{15v^5} = \frac{3}{5v^4}$$

$$\frac{16q^2}{20q^3} = \frac{4}{5q}$$

$$\frac{15d^5}{9d} = \frac{5d^4}{3}$$

$$\frac{20w^5}{50w^5} = \frac{2}{5}$$

$$\frac{50y^2}{20y^4} = \frac{5}{2y^2}$$

$$\frac{8u^4}{20u^5} = \frac{2}{5u}$$

Simplify each expression.

$\frac{9q^4}{3q^5} =$

$\frac{30f^5}{40f^4} =$

$\frac{16d^6}{20d^4} =$

$\frac{25x^2}{15x} =$

$\frac{3m^6}{15m^2} =$

$\frac{50g^3}{40g^3} =$

$\frac{10u^4}{15u^4} =$

$\frac{30w^3}{50w^2} =$

$\frac{8w}{4w^2} =$

$\frac{15y^5}{25y^6} =$

$\frac{12k}{9k^6} =$

$\frac{20k^2}{30k} =$

$\frac{4x^3}{8x^2} =$

$\frac{4a^3}{40a} =$

$\frac{12v^4}{4v^2} =$

$\frac{15q^6}{6q^5} =$

$\frac{25h^3}{5h^3} =$

$\frac{3x^2}{9x^3} =$

$\frac{16y^5}{12y^4} =$

$\frac{8q^2}{12q} =$

# Answer Key

Simplify each expression.

$\frac{9q^4}{3q^5} = \frac{3}{q}$

$\frac{30f^5}{40f^4} = \frac{3f}{4}$

$\frac{16d^6}{20d^4} = \frac{4d^2}{5}$

$\frac{25x^2}{15x} = \frac{5x}{3}$

$\frac{3m^6}{15m^2} = \frac{m^4}{5}$

$\frac{50g^3}{40g^3} = \frac{5}{4}$

$\frac{10u^4}{15u^4} = \frac{2}{3}$

$\frac{30w^3}{50w^2} = \frac{3w}{5}$

$\frac{8w}{4w^2} = \frac{2}{w}$

$\frac{15y^5}{25y^6} = \frac{3}{5y}$

$\frac{12k}{9k^6} = \frac{4}{3k^5}$

$\frac{20k^2}{30k} = \frac{2k}{3}$

$\frac{4x^3}{8x^2} = \frac{x}{2}$

$\frac{4a^3}{40a} = \frac{a^2}{10}$

$\frac{12v^4}{4v^2} = 3v^2$

$\frac{15q^6}{6q^5} = \frac{5q}{2}$

$\frac{25h^3}{5h^3} = 5$

$\frac{3x^2}{9x^3} = \frac{1}{3x}$

$\frac{16y^5}{12y^4} = \frac{4y}{3}$

$\frac{8q^2}{12q} = \frac{2q}{3}$

Simplify each expression.

$\frac{8r^5}{20r^5} =$

$\frac{12t^6}{15t} =$

$\frac{15r^5}{10r^4} =$

$\frac{15q^6}{25q^4} =$

$\frac{9p^6}{6p^3} =$

$\frac{2p^6}{4p^2} =$

$\frac{6k}{15k^2} =$

$\frac{6r^4}{8r^2} =$

$\frac{10b^6}{100b^3} =$

$\frac{20q^3}{10q^2} =$

$\frac{15x}{25x} =$

$\frac{5q^2}{25q^4} =$

$\frac{8g^3}{20g^2} =$

$\frac{6y^6}{2y^3} =$

$\frac{20y}{16y^5} =$

$\frac{4b}{6b^4} =$

$\frac{12a^5}{4a^5} =$

$\frac{40f}{50f^4} =$

$\frac{30g^3}{50g^4} =$

$\frac{6w}{15w^3} =$

# Answer Key

Simplify each expression.

$\frac{8r^5}{20r^5} = \frac{2}{5}$

$\frac{12t^6}{15t} = \frac{4t^5}{5}$

$\frac{15r^5}{10r^4} = \frac{3r}{2}$

$\frac{15q^6}{25q^4} = \frac{3q^2}{5}$

$\frac{9p^6}{6p^3} = \frac{3p^3}{2}$

$\frac{2p^6}{4p^2} = \frac{p^4}{2}$

$\frac{6k}{15k^2} = \frac{2}{5k}$

$\frac{6r^4}{8r^2} = \frac{3r^2}{4}$

$\frac{10b^6}{100b^3} = \frac{b^3}{10}$

$\frac{20q^3}{10q^2} = 2q$

$\frac{15x}{25x} = \frac{3}{5}$

$\frac{5q^2}{25q^4} = \frac{1}{5q^2}$

$\frac{8g^3}{20g^2} = \frac{2g}{5}$

$\frac{6y^6}{2y^3} = 3y^3$

$\frac{20y}{16y^5} = \frac{5}{4y^4}$

$\frac{4b}{6b^4} = \frac{2}{3b^3}$

$\frac{12a^5}{4a^5} = 3$

$\frac{40f}{50f^4} = \frac{4}{5f^3}$

$\frac{30g^3}{50g^4} = \frac{3}{5g}$

$\frac{6w}{15w^3} = \frac{2}{5w^2}$

Simplify each expression.

$\frac{12r^2}{8r^3} =$

$\frac{4h}{8h^5} =$

$\frac{6y^5}{4y^6} =$

$\frac{15x^4}{12x^3} =$

$\frac{2x^4}{10x} =$

$\frac{4w^2}{40w^3} =$

$\frac{30g}{50g^5} =$

$\frac{6w^3}{8w^4} =$

$\frac{10g}{40g} =$

$\frac{15q^3}{5q} =$

$\frac{5q^4}{15q} =$

$\frac{20f}{12f^2} =$

$\frac{12g^2}{16g^3} =$

$\frac{6b^5}{4b^5} =$

$\frac{30g^3}{40g^2} =$

$\frac{4r^4}{12r^5} =$

$\frac{20u}{8u^2} =$

$\frac{15g^4}{12g} =$

$\frac{2h^4}{10h^3} =$

$\frac{30w^3}{50w^2} =$

# Answer Key

Simplify each expression.

$$\frac{12r^2}{8r^3} = \frac{3}{2r}$$

$$\frac{5q^4}{15q} = \frac{q^3}{3}$$

$$\frac{4h}{8h^5} = \frac{1}{2h^4}$$

$$\frac{20f}{12f^2} = \frac{5}{3f}$$

$$\frac{6y^5}{4y^6} = \frac{3}{2y}$$

$$\frac{12g^2}{16g^3} = \frac{3}{4g}$$

$$\frac{15x^4}{12x^3} = \frac{5x}{4}$$

$$\frac{6b^5}{4b^5} = \frac{3}{2}$$

$$\frac{2x^4}{10x} = \frac{x^3}{5}$$

$$\frac{30g^3}{40g^2} = \frac{3g}{4}$$

$$\frac{4w^2}{40w^3} = \frac{1}{10w}$$

$$\frac{4r^4}{12r^5} = \frac{1}{3r}$$

$$\frac{30g}{50g^5} = \frac{3}{5g^4}$$

$$\frac{20u}{8u^2} = \frac{5}{2u}$$

$$\frac{6w^3}{8w^4} = \frac{3}{4w}$$

$$\frac{15g^4}{12g} = \frac{5g^3}{4}$$

$$\frac{10g}{40g} = \frac{1}{4}$$

$$\frac{2h^4}{10h^3} = \frac{h}{5}$$

$$\frac{15q^3}{5q} = 3q^2$$

$$\frac{30w^3}{50w^2} = \frac{3w}{5}$$

Simplify each expression.

$\dfrac{10p^3}{4p^2} =$

$\dfrac{2m}{8m^5} =$

$\dfrac{20r^6}{50r^6} =$

$\dfrac{6x}{15x^4} =$

$\dfrac{2w^5}{8w^3} =$

$\dfrac{3q^3}{6q^4} =$

$\dfrac{10t}{6t^4} =$

$\dfrac{50n^2}{40n^6} =$

$\dfrac{12w^5}{3w} =$

$\dfrac{50b^3}{30b^2} =$

$\dfrac{20x^4}{4x^3} =$

$\dfrac{20y}{5y^3} =$

$\dfrac{6f^2}{10f^5} =$

$\dfrac{6y^5}{8y^3} =$

$\dfrac{4v^3}{40v^2} =$

$\dfrac{40w^2}{4w^4} =$

$\dfrac{6y^6}{3y^2} =$

$\dfrac{10v}{25v^5} =$

$\dfrac{6w^4}{4w^6} =$

$\dfrac{10x^2}{15x^5} =$

# Answer Key

Simplify each expression.

$$\frac{10p^3}{4p^2} = \frac{5p}{2}$$

$$\frac{2m}{8m^5} = \frac{1}{4m^4}$$

$$\frac{20r^6}{50r^6} = \frac{2}{5}$$

$$\frac{6x}{15x^4} = \frac{2}{5x^3}$$

$$\frac{2w^5}{8w^3} = \frac{w^2}{4}$$

$$\frac{3q^3}{6q^4} = \frac{1}{2q}$$

$$\frac{10t}{6t^4} = \frac{5}{3t^3}$$

$$\frac{50n^2}{40n^6} = \frac{5}{4n^4}$$

$$\frac{12w^5}{3w} = 4w^4$$

$$\frac{50b^3}{30b^2} = \frac{5b}{3}$$

$$\frac{20x^4}{4x^3} = 5x$$

$$\frac{20y}{5y^3} = \frac{4}{y^2}$$

$$\frac{6f^2}{10f^5} = \frac{3}{5f^3}$$

$$\frac{6y^5}{8y^3} = \frac{3y^2}{4}$$

$$\frac{4v^3}{40v^2} = \frac{v}{10}$$

$$\frac{40w^2}{4w^4} = \frac{10}{w^2}$$

$$\frac{6y^6}{3y^2} = 2y^4$$

$$\frac{10v}{25v^5} = \frac{2}{5v^4}$$

$$\frac{6w^4}{4w^6} = \frac{3}{2w^2}$$

$$\frac{10x^2}{15x^5} = \frac{2}{3x^3}$$

Simplify each expression.

$\frac{4v^5}{12v^3} =$

$\frac{3h^4}{9h^2} =$

$\frac{9y^4}{3y^3} =$

$\frac{3g}{12g} =$

$\frac{20y^2}{50y^6} =$

$\frac{8v^3}{20v^5} =$

$\frac{25p^6}{20p^5} =$

$\frac{12v^2}{3v^4} =$

$\frac{3h^4}{12h^5} =$

$\frac{9b^2}{15b^2} =$

$\frac{100h}{10h^4} =$

$\frac{2t^2}{20t^5} =$

$\frac{9w^5}{15w} =$

$\frac{4r^6}{20r^6} =$

$\frac{15d^4}{6d^5} =$

$\frac{2b^4}{6b^2} =$

$\frac{5m^2}{20m^3} =$

$\frac{4b^5}{12b^6} =$

$\frac{25v^3}{20v^2} =$

$\frac{30a^4}{50a^3} =$

# Answer Key

Simplify each expression.

$$\frac{4v^5}{12v^3} = \frac{v^2}{3}$$

$$\frac{100h}{10h^4} = \frac{10}{h^3}$$

$$\frac{3h^4}{9h^2} = \frac{h^2}{3}$$

$$\frac{2t^2}{20t^5} = \frac{1}{10t^3}$$

$$\frac{9y^4}{3y^3} = 3y$$

$$\frac{9w^5}{15w} = \frac{3w^4}{5}$$

$$\frac{3g}{12g} = \frac{1}{4}$$

$$\frac{4r^6}{20r^6} = \frac{1}{5}$$

$$\frac{20y^2}{50y^6} = \frac{2}{5y^4}$$

$$\frac{15d^4}{6d^5} = \frac{5}{2d}$$

$$\frac{8v^3}{20v^5} = \frac{2}{5v^2}$$

$$\frac{2b^4}{6b^2} = \frac{b^2}{3}$$

$$\frac{25p^6}{20p^5} = \frac{5p}{4}$$

$$\frac{5m^2}{20m^3} = \frac{1}{4m}$$

$$\frac{12v^2}{3v^4} = \frac{4}{v^2}$$

$$\frac{4b^5}{12b^6} = \frac{1}{3b}$$

$$\frac{3h^4}{12h^5} = \frac{1}{4h}$$

$$\frac{25v^3}{20v^2} = \frac{5v}{4}$$

$$\frac{9b^2}{15b^2} = \frac{3}{5}$$

$$\frac{30a^4}{50a^3} = \frac{3a}{5}$$

Simplify each expression.

$\frac{50p^2}{10p^6} =$

$\frac{9m^5}{3m^3} =$

$\frac{8m^4}{20m^6} =$

$\frac{10u^5}{20u^3} =$

$\frac{20v}{8v^6} =$

$\frac{5h^2}{20h} =$

$\frac{10q^2}{4q} =$

$\frac{16g}{12g^3} =$

$\frac{12m^2}{3m^3} =$

$\frac{6d^2}{8d^5} =$

$\frac{12v^2}{9v^5} =$

$\frac{6n^2}{3n^6} =$

$\frac{16k^5}{4k} =$

$\frac{8u^6}{10u^2} =$

$\frac{30q}{10q^3} =$

$\frac{15q^6}{10q^4} =$

$\frac{4m^6}{16m^2} =$

$\frac{16v}{4v^6} =$

$\frac{8r^6}{12r^2} =$

$\frac{5v^2}{15v^6} =$

# Answer Key

Simplify each expression.

$$\frac{50p^2}{10p^6} = \frac{5}{p^4}$$

$$\frac{9m^5}{3m^3} = 3m^2$$

$$\frac{8m^4}{20m^6} = \frac{2}{5m^2}$$

$$\frac{10u^5}{20u^3} = \frac{u^2}{2}$$

$$\frac{20v}{8v^6} = \frac{5}{2v^5}$$

$$\frac{5h^2}{20h} = \frac{h}{4}$$

$$\frac{10q^2}{4q} = \frac{5q}{2}$$

$$\frac{16g}{12g^3} = \frac{4}{3g^2}$$

$$\frac{12m^2}{3m^3} = \frac{4}{m}$$

$$\frac{6d^2}{8d^5} = \frac{3}{4d^3}$$

$$\frac{12v^2}{9v^5} = \frac{4}{3v^3}$$

$$\frac{6n^2}{3n^6} = \frac{2}{n^4}$$

$$\frac{16k^5}{4k} = 4k^4$$

$$\frac{8u^6}{10u^2} = \frac{4u^4}{5}$$

$$\frac{30q}{10q^3} = \frac{3}{q^2}$$

$$\frac{15q^6}{10q^4} = \frac{3q^2}{2}$$

$$\frac{4m^6}{16m^2} = \frac{m^4}{4}$$

$$\frac{16v}{4v^6} = \frac{4}{v^5}$$

$$\frac{8r^6}{12r^2} = \frac{2r^4}{3}$$

$$\frac{5v^2}{15v^6} = \frac{1}{3v^4}$$

Simplify each expression.

$\frac{6w^5}{3w^4} =$

$\frac{12h^3}{20h^4} =$

$\frac{2g^2}{10g^2} =$

$\frac{20f^6}{2f^2} =$

$\frac{4n^3}{10n^4} =$

$\frac{20h^4}{8h^3} =$

$\frac{3x^4}{12x^6} =$

$\frac{4w^5}{10w^3} =$

$\frac{9a^2}{15a^6} =$

$\frac{10h^6}{15h^6} =$

$\frac{50r^5}{10r^3} =$

$\frac{12g^2}{16g^5} =$

$\frac{10k^6}{8k^4} =$

$\frac{25m^4}{10m^2} =$

$\frac{4g^6}{16g^3} =$

$\frac{3m^2}{12m} =$

$\frac{5q^5}{50q^2} =$

$\frac{8t}{20t^5} =$

$\frac{30a^2}{20a^5} =$

$\frac{2d}{8d^3} =$

# Answer Key

Simplify each expression.

$$\frac{6w^5}{3w^4} = 2w$$

$$\frac{12h^3}{20h^4} = \frac{3}{5h}$$

$$\frac{2g^2}{10g^2} = \frac{1}{5}$$

$$\frac{20f^6}{2f^2} = 10f^4$$

$$\frac{4n^3}{10n^4} = \frac{2}{5n}$$

$$\frac{20h^4}{8h^3} = \frac{5h}{2}$$

$$\frac{3x^4}{12x^6} = \frac{1}{4x^2}$$

$$\frac{4w^5}{10w^3} = \frac{2w^2}{5}$$

$$\frac{9a^2}{15a^6} = \frac{3}{5a^4}$$

$$\frac{10h^6}{15h^6} = \frac{2}{3}$$

$$\frac{50r^5}{10r^3} = 5r^2$$

$$\frac{12g^2}{16g^5} = \frac{3}{4g^3}$$

$$\frac{10k^6}{8k^4} = \frac{5k^2}{4}$$

$$\frac{25m^4}{10m^2} = \frac{5m^2}{2}$$

$$\frac{4g^6}{16g^3} = \frac{g^3}{4}$$

$$\frac{3m^2}{12m} = \frac{m}{4}$$

$$\frac{5q^5}{50q^2} = \frac{q^3}{10}$$

$$\frac{8t}{20t^5} = \frac{2}{5t^4}$$

$$\frac{30a^2}{20a^5} = \frac{3}{2a^3}$$

$$\frac{2d}{8d^3} = \frac{1}{4d^2}$$

Simplify each expression.

$\frac{20f^2}{10f^5} =$

$\frac{5r}{20r^2} =$

$\frac{9d^4}{3d^3} =$

$\frac{25h^3}{5h^6} =$

$\frac{3p^3}{15p^4} =$

$\frac{4m^2}{12m^3} =$

$\frac{5b^2}{15b^2} =$

$\frac{2f^3}{6f^3} =$

$\frac{5g^2}{15g^2} =$

$\frac{8v^5}{10v^4} =$

$\frac{40u^3}{30u^4} =$

$\frac{12t^2}{3t^2} =$

$\frac{15b^2}{25b^3} =$

$\frac{100h^2}{10h^3} =$

$\frac{5n^3}{50n^2} =$

$\frac{8p^2}{2p^2} =$

$\frac{6r^3}{4r} =$

$\frac{2q^4}{6q^2} =$

$\frac{9t^5}{6t} =$

$\frac{4f^2}{10f^2} =$

# Answer Key

Simplify each expression.

$\frac{20f^2}{10f^5} = \frac{2}{f^3}$

$\frac{5r}{20r^2} = \frac{1}{4r}$

$\frac{9d^4}{3d^3} = 3d$

$\frac{25h^3}{5h^6} = \frac{5}{h^3}$

$\frac{3p^3}{15p^4} = \frac{1}{5p}$

$\frac{4m^2}{12m^3} = \frac{1}{3m}$

$\frac{5b^2}{15b^2} = \frac{1}{3}$

$\frac{2f^3}{6f^3} = \frac{1}{3}$

$\frac{5g^2}{15g^2} = \frac{1}{3}$

$\frac{8v^5}{10v^4} = \frac{4v}{5}$

$\frac{40u^3}{30u^4} = \frac{4}{3u}$

$\frac{12t^2}{3t^2} = 4$

$\frac{15b^2}{25b^3} = \frac{3}{5b}$

$\frac{100h^2}{10h^3} = \frac{10}{h}$

$\frac{5n^3}{50n^2} = \frac{n}{10}$

$\frac{8p^2}{2p^2} = 4$

$\frac{6r^3}{4r} = \frac{3r^2}{2}$

$\frac{2q^4}{6q^2} = \frac{q^2}{3}$

$\frac{9t^5}{6t} = \frac{3t^4}{2}$

$\frac{4f^2}{10f^2} = \frac{2}{5}$

Simplify each expression.

$\frac{2v^2}{20v^2} =$

$\frac{6x^6}{2x^3} =$

$\frac{12g}{16g^6} =$

$\frac{8b^5}{10b^5} =$

$\frac{8n^3}{4n^2} =$

$\frac{25v^3}{20v} =$

$\frac{50u}{40u^3} =$

$\frac{8n^2}{20n^3} =$

$\frac{15x^4}{10x^4} =$

$\frac{6a^4}{2a^5} =$

$\frac{5y}{25y^5} =$

$\frac{15r}{6r^2} =$

$\frac{12m^2}{9m} =$

$\frac{16n^3}{20n^6} =$

$\frac{12m^3}{8m^2} =$

$\frac{4u^2}{10u^4} =$

$\frac{3w}{15w^4} =$

$\frac{2h^2}{4h^2} =$

$\frac{100d^4}{10d^4} =$

$\frac{10m^3}{25m^2} =$

# Answer Key

Simplify each expression.

$$\frac{2v^2}{20v^2} = \frac{1}{10}$$

$$\frac{6x^6}{2x^3} = 3x^3$$

$$\frac{12g}{16g^6} = \frac{3}{4g^5}$$

$$\frac{8b^5}{10b^5} = \frac{4}{5}$$

$$\frac{8n^3}{4n^2} = 2n$$

$$\frac{25v^3}{20v} = \frac{5v^2}{4}$$

$$\frac{50u}{40u^3} = \frac{5}{4u^2}$$

$$\frac{8n^2}{20n^3} = \frac{2}{5n}$$

$$\frac{15x^4}{10x^4} = \frac{3}{2}$$

$$\frac{6a^4}{2a^5} = \frac{3}{a}$$

$$\frac{5y}{25y^5} = \frac{1}{5y^4}$$

$$\frac{15r}{6r^2} = \frac{5}{2r}$$

$$\frac{12m^2}{9m} = \frac{4m}{3}$$

$$\frac{16n^3}{20n^6} = \frac{4}{5n^3}$$

$$\frac{12m^3}{8m^2} = \frac{3m}{2}$$

$$\frac{4u^2}{10u^4} = \frac{2}{5u^2}$$

$$\frac{3w}{15w^4} = \frac{1}{5w^3}$$

$$\frac{2h^2}{4h^2} = \frac{1}{2}$$

$$\frac{100d^4}{10d^4} = 10$$

$$\frac{10m^3}{25m^2} = \frac{2m}{5}$$

Simplify each expression.

$\frac{5x^2}{25x^4} =$

$\frac{40b}{30b} =$

$\frac{15y^2}{25y^4} =$

$\frac{6p^4}{10p^3} =$

$\frac{40f^2}{30f^3} =$

$\frac{25k^5}{15k^4} =$

$\frac{2r}{4r^3} =$

$\frac{100b^2}{10b^3} =$

$\frac{6t^3}{15t^6} =$

$\frac{15p^5}{5p^5} =$

$\frac{12t^2}{15t^3} =$

$\frac{4n^2}{12n^4} =$

$\frac{6m^5}{15m^6} =$

$\frac{100n^2}{10n^2} =$

$\frac{12b^2}{15b^5} =$

$\frac{3g^6}{30g^6} =$

$\frac{10f^3}{2f^6} =$

$\frac{5x}{15x^2} =$

$\frac{20v^6}{2v^4} =$

$\frac{2n^5}{6n^4} =$

# Answer Key

Simplify each expression.

$$\frac{5x^2}{25x^4} = \frac{1}{5x^2}$$

$$\frac{40b}{30b} = \frac{4}{3}$$

$$\frac{15y^2}{25y^4} = \frac{3}{5y^2}$$

$$\frac{6p^4}{10p^3} = \frac{3p}{5}$$

$$\frac{40f^2}{30f^3} = \frac{4}{3f}$$

$$\frac{25k^5}{15k^4} = \frac{5k}{3}$$

$$\frac{2r}{4r^3} = \frac{1}{2r^2}$$

$$\frac{100b^2}{10b^3} = \frac{10}{b}$$

$$\frac{6t^3}{15t^6} = \frac{2}{5t^3}$$

$$\frac{15p^5}{5p^5} = 3$$

$$\frac{12t^2}{15t^3} = \frac{4}{5t}$$

$$\frac{4n^2}{12n^4} = \frac{1}{3n^2}$$

$$\frac{6m^5}{15m^6} = \frac{2}{5m}$$

$$\frac{100n^2}{10n^2} = 10$$

$$\frac{12b^2}{15b^5} = \frac{4}{5b^3}$$

$$\frac{3g^6}{30g^6} = \frac{1}{10}$$

$$\frac{10f^3}{2f^6} = \frac{5}{f^3}$$

$$\frac{5x}{15x^2} = \frac{1}{3x}$$

$$\frac{20v^6}{2v^4} = 10v^2$$

$$\frac{2n^5}{6n^4} = \frac{n}{3}$$

Simplify each expression.

$\frac{30k}{10k^5} =$

$\frac{30a^2}{20a^3} =$

$\frac{20q^3}{50q} =$

$\frac{25f}{10f^5} =$

$\frac{2a^2}{10a} =$

$\frac{15n^3}{25n^5} =$

$\frac{10x}{40x^3} =$

$\frac{30h^4}{50h^4} =$

$\frac{20f^2}{50f^5} =$

$\frac{8n^2}{12n^2} =$

$\frac{12r^2}{4r} =$

$\frac{10y^5}{5y^4} =$

$\frac{40a^6}{50a^5} =$

$\frac{20m^4}{5m^5} =$

$\frac{30b}{3b^4} =$

$\frac{12g^5}{4g^4} =$

$\frac{12m^3}{8m^3} =$

$\frac{10k^2}{50k} =$

$\frac{20a^4}{15a^4} =$

$\frac{12u^3}{3u^3} =$

# Answer Key

Simplify each expression.

$$\frac{30k}{10k^5} = \frac{3}{k^4}$$

$$\frac{30a^2}{20a^3} = \frac{3}{2a}$$

$$\frac{20q^3}{50q} = \frac{2q^2}{5}$$

$$\frac{25f}{10f^5} = \frac{5}{2f^4}$$

$$\frac{2a^2}{10a} = \frac{a}{5}$$

$$\frac{15n^3}{25n^5} = \frac{3}{5n^2}$$

$$\frac{10x}{40x^3} = \frac{1}{4x^2}$$

$$\frac{30h^4}{50h^4} = \frac{3}{5}$$

$$\frac{20f^2}{50f^5} = \frac{2}{5f^3}$$

$$\frac{8n^2}{12n^2} = \frac{2}{3}$$

$$\frac{12r^2}{4r} = 3r$$

$$\frac{10y^5}{5y^4} = 2y$$

$$\frac{40a^6}{50a^5} = \frac{4a}{5}$$

$$\frac{20m^4}{5m^5} = \frac{4}{m}$$

$$\frac{30b}{3b^4} = \frac{10}{b^3}$$

$$\frac{12g^5}{4g^4} = 3g$$

$$\frac{12m^3}{8m^3} = \frac{3}{2}$$

$$\frac{10k^2}{50k} = \frac{k}{5}$$

$$\frac{20a^4}{15a^4} = \frac{4}{3}$$

$$\frac{12u^3}{3u^3} = 4$$

Simplify each expression.

$$\frac{20v^3}{2v^3} =$$

$$\frac{20f^5}{25f^5} =$$

$$\frac{30v^6}{20v^6} =$$

$$\frac{4a^3}{16a^5} =$$

$$\frac{25n^2}{10n^3} =$$

$$\frac{12h^3}{8h^2} =$$

$$\frac{10x}{100x} =$$

$$\frac{4p^4}{10p} =$$

$$\frac{30k}{20k^5} =$$

$$\frac{8m^6}{20m^2} =$$

$$\frac{20a^5}{5a^5} =$$

$$\frac{12b^2}{20b^6} =$$

$$\frac{16h^6}{20h^6} =$$

$$\frac{15k}{5k^4} =$$

$$\frac{10g^4}{15g^5} =$$

$$\frac{6v^2}{8v^2} =$$

$$\frac{16m^6}{4m^5} =$$

$$\frac{20n^2}{8n^2} =$$

$$\frac{9r^3}{15r^5} =$$

$$\frac{9k^3}{3k^6} =$$

# Answer Key

Simplify each expression.

$$\frac{20v^3}{2v^3} = 10$$

$$\frac{20a^5}{5a^5} = 4$$

$$\frac{20f^5}{25f^5} = \frac{4}{5}$$

$$\frac{12b^2}{20b^6} = \frac{3}{5b^4}$$

$$\frac{30v^6}{20v^6} = \frac{3}{2}$$

$$\frac{16h^6}{20h^6} = \frac{4}{5}$$

$$\frac{4a^3}{16a^5} = \frac{1}{4a^2}$$

$$\frac{15k}{5k^4} = \frac{3}{k^3}$$

$$\frac{25n^2}{10n^3} = \frac{5}{2n}$$

$$\frac{10g^4}{15g^5} = \frac{2}{3g}$$

$$\frac{12h^3}{8h^2} = \frac{3h}{2}$$

$$\frac{6v^2}{8v^2} = \frac{3}{4}$$

$$\frac{10x}{100x} = \frac{1}{10}$$

$$\frac{16m^6}{4m^5} = 4m$$

$$\frac{4p^4}{10p} = \frac{2p^3}{5}$$

$$\frac{20n^2}{8n^2} = \frac{5}{2}$$

$$\frac{30k}{20k^5} = \frac{3}{2k^4}$$

$$\frac{9r^3}{15r^5} = \frac{3}{5r^2}$$

$$\frac{8m^6}{20m^2} = \frac{2m^4}{5}$$

$$\frac{9k^3}{3k^6} = \frac{3}{k^3}$$

Simplify each expression.

$\frac{10t^3}{4t^5} =$

$\frac{2n}{20n^3} =$

$\frac{2a^2}{20a^5} =$

$\frac{10h^6}{40h^2} =$

$\frac{30n^2}{50n} =$

$\frac{50h^2}{40h^5} =$

$\frac{30n^4}{10n^6} =$

$\frac{6g^4}{9g^6} =$

$\frac{6x}{10x^6} =$

$\frac{30b^6}{40b^2} =$

$\frac{8t}{2t^3} =$

$\frac{4k^2}{12k^5} =$

$\frac{15q^6}{9q} =$

$\frac{20a}{25a^5} =$

$\frac{10k^5}{40k^6} =$

$\frac{25f^2}{10f^3} =$

$\frac{8d^2}{20d^2} =$

$\frac{2h^2}{8h^5} =$

$\frac{20q^6}{4q^6} =$

$\frac{12q}{9q^6} =$

# Answer Key

Simplify each expression.

$$\frac{10t^3}{4t^5} = \frac{5}{2t^2}$$

$$\frac{2n}{20n^3} = \frac{1}{10n^2}$$

$$\frac{2a^2}{20a^5} = \frac{1}{10a^3}$$

$$\frac{10h^6}{40h^2} = \frac{h^4}{4}$$

$$\frac{30n^2}{50n} = \frac{3n}{5}$$

$$\frac{50h^2}{40h^5} = \frac{5}{4h^3}$$

$$\frac{30n^4}{10n^6} = \frac{3}{n^2}$$

$$\frac{6g^4}{9g^6} = \frac{2}{3g^2}$$

$$\frac{6x}{10x^6} = \frac{3}{5x^5}$$

$$\frac{30b^6}{40b^2} = \frac{3b^4}{4}$$

$$\frac{8t}{2t^3} = \frac{4}{t^2}$$

$$\frac{4k^2}{12k^5} = \frac{1}{3k^3}$$

$$\frac{15q^6}{9q} = \frac{5q^5}{3}$$

$$\frac{20a}{25a^5} = \frac{4}{5a^4}$$

$$\frac{10k^5}{40k^6} = \frac{1}{4k}$$

$$\frac{25f^2}{10f^3} = \frac{5}{2f}$$

$$\frac{8d^2}{20d^2} = \frac{2}{5}$$

$$\frac{2h^2}{8h^5} = \frac{1}{4h^3}$$

$$\frac{20q^6}{4q^6} = 5$$

$$\frac{12q}{9q^6} = \frac{4}{3q^5}$$

Simplify each expression.

$\frac{9h^3}{15h^6} =$

$\frac{40n}{50n^4} =$

$\frac{12p^6}{20p^2} =$

$\frac{10b^6}{30b^2} =$

$\frac{16g^4}{12g^5} =$

$\frac{10d^2}{25d^5} =$

$\frac{50x}{10x^5} =$

$\frac{50d^5}{30d^2} =$

$\frac{8p}{4p^3} =$

$\frac{20b^4}{5b^2} =$

$\frac{40n^6}{50n^3} =$

$\frac{9k^2}{12k^3} =$

$\frac{16a^2}{12a^4} =$

$\frac{12u^5}{16u^5} =$

$\frac{5f^3}{15f^6} =$

$\frac{4u^2}{2u^3} =$

$\frac{40b^3}{4b^3} =$

$\frac{20r^6}{5r^2} =$

$\frac{4h^5}{8h^4} =$

$\frac{4a}{40a} =$

# Answer Key

Simplify each expression.

$$\frac{9h^3}{15h^6} = \frac{3}{5h^3}$$

$$\frac{40n}{50n^4} = \frac{4}{5n^3}$$

$$\frac{12p^6}{20p^2} = \frac{3p^4}{5}$$

$$\frac{10b^6}{30b^2} = \frac{b^4}{3}$$

$$\frac{16g^4}{12g^5} = \frac{4}{3g}$$

$$\frac{10d^2}{25d^5} = \frac{2}{5d^3}$$

$$\frac{50x}{10x^5} = \frac{5}{x^4}$$

$$\frac{50d^5}{30d^2} = \frac{5d^3}{3}$$

$$\frac{8p}{4p^3} = \frac{2}{p^2}$$

$$\frac{20b^4}{5b^2} = 4b^2$$

$$\frac{40n^6}{50n^3} = \frac{4n^3}{5}$$

$$\frac{9k^2}{12k^3} = \frac{3}{4k}$$

$$\frac{16a^2}{12a^4} = \frac{4}{3a^2}$$

$$\frac{12u^5}{16u^5} = \frac{3}{4}$$

$$\frac{5f^3}{15f^6} = \frac{1}{3f^3}$$

$$\frac{4u^2}{2u^3} = \frac{2}{u}$$

$$\frac{40b^3}{4b^3} = 10$$

$$\frac{20r^6}{5r^2} = 4r^4$$

$$\frac{4h^5}{8h^4} = \frac{h}{2}$$

$$\frac{4a}{40a} = \frac{1}{10}$$

Simplify each expression.

$\frac{5v^6}{20v^5} =$

$\frac{30d^2}{40d} =$

$\frac{3x^5}{6x^3} =$

$\frac{5m^5}{10m^6} =$

$\frac{12p^2}{4p^5} =$

$\frac{10x^3}{30x^4} =$

$\frac{15r}{25r^5} =$

$\frac{8b^4}{12b^4} =$

$\frac{25d^6}{10d^4} =$

$\frac{25u^5}{15u^3} =$

$\frac{9v^5}{3v} =$

$\frac{6f^3}{4f^2} =$

$\frac{40w^5}{4w^4} =$

$\frac{8t^4}{4t} =$

$\frac{8b^2}{10b^5} =$

$\frac{25w}{10w^5} =$

$\frac{30n^5}{20n^3} =$

$\frac{6q^3}{2q^2} =$

$\frac{2b^5}{10b^2} =$

$\frac{4d}{16d^2} =$

# Answer Key

Simplify each expression.

$$\frac{5v^6}{20v^5} = \frac{v}{4}$$

$$\frac{30d^2}{40d} = \frac{3d}{4}$$

$$\frac{3x^5}{6x^3} = \frac{x^2}{2}$$

$$\frac{5m^5}{10m^6} = \frac{1}{2m}$$

$$\frac{12p^2}{4p^5} = \frac{3}{p^3}$$

$$\frac{10x^3}{30x^4} = \frac{1}{3x}$$

$$\frac{15r}{25r^5} = \frac{3}{5r^4}$$

$$\frac{8b^4}{12b^4} = \frac{2}{3}$$

$$\frac{25d^6}{10d^4} = \frac{5d^2}{2}$$

$$\frac{25u^5}{15u^3} = \frac{5u^2}{3}$$

$$\frac{9v^5}{3v} = 3v^4$$

$$\frac{6f^3}{4f^2} = \frac{3f}{2}$$

$$\frac{40w^5}{4w^4} = 10w$$

$$\frac{8t^4}{4t} = 2t^3$$

$$\frac{8b^2}{10b^5} = \frac{4}{5b^3}$$

$$\frac{25w}{10w^5} = \frac{5}{2w^4}$$

$$\frac{30n^5}{20n^3} = \frac{3n^2}{2}$$

$$\frac{6q^3}{2q^2} = 3q$$

$$\frac{2b^5}{10b^2} = \frac{b^3}{5}$$

$$\frac{4d}{16d^2} = \frac{1}{4d}$$

Simplify each expression.

$\frac{6rw^2}{10rw^2} =$

$\frac{20w^2u^5}{10w^2u} =$

$\frac{10r^4u^6}{25r^3u^2} =$

$\frac{10n^3y^5}{30n^3y^4} =$

$\frac{4ay^2}{16a^6y} =$

$\frac{10y^4x^3}{30yx} =$

$\frac{40a^2v^6}{50a^2v} =$

$\frac{25x^4k^3}{10x^4k} =$

$\frac{3a^5q^3}{30a^4q^4} =$

$\frac{100h^3t^6}{10h^3t} =$

$\frac{20y^4u^2}{10y^6u^4} =$

$\frac{8a^5m^2}{6a^5m^3} =$

$\frac{2k^3a^2}{20k^2a^2} =$

$\frac{4g^5b^5}{10gb} =$

$\frac{4r^3n}{10r^5n^2} =$

$\frac{20dp^5}{4dp} =$

$\frac{12g^4d^4}{15g^5d^5} =$

$\frac{6tk^2}{2t^6k^5} =$

$\frac{20v^2p^4}{16v^3p^6} =$

$\frac{4w^5q}{10wq^4} =$

# Answer Key

Simplify each expression.

$\frac{6rw^2}{10rw^2} = \frac{3}{5}$

$\frac{20w^2u^5}{10w^2u} = 2u^4$

$\frac{10r^4u^6}{25r^3u^2} = \frac{2ru^4}{5}$

$\frac{10n^3y^5}{30n^3y^4} = \frac{y}{3}$

$\frac{4ay^2}{16a^6y} = \frac{y}{4a^5}$

$\frac{10y^4x^3}{30yx} = \frac{y^3x^2}{3}$

$\frac{40a^2v^6}{50a^2v} = \frac{4v^5}{5}$

$\frac{25x^4k^3}{10x^4k} = \frac{5k^2}{2}$

$\frac{3a^5q^3}{30a^4q^4} = \frac{a}{10q}$

$\frac{100h^3t^6}{10h^3t} = 10t^5$

$\frac{20y^4u^2}{10y^6u^4} = \frac{2}{y^2u^2}$

$\frac{8a^5m^2}{6a^5m^3} = \frac{4}{3m}$

$\frac{2k^3a^2}{20k^2a^2} = \frac{k}{10}$

$\frac{4g^5b^5}{10gb} = \frac{2g^4b^4}{5}$

$\frac{4r^3n}{10r^5n^2} = \frac{2}{5r^2n}$

$\frac{20dp^5}{4dp} = 5p^4$

$\frac{12g^4d^4}{15g^5d^5} = \frac{4}{5gd}$

$\frac{6tk^2}{2t^6k^5} = \frac{3}{t^5k^3}$

$\frac{20v^2p^4}{16v^3p^6} = \frac{5}{4vp^2}$

$\frac{4w^5q}{10wq^4} = \frac{2w^4}{5q^3}$

Simplify each expression.

$\frac{50u^5a}{40u^2a^5} =$

$\frac{10k^5m^2}{6km^5} =$

$\frac{9q^2g^6}{12q^4g^4} =$

$\frac{20t^2m^5}{16t^2m^3} =$

$\frac{8w^6a^6}{4wa^2} =$

$\frac{5n^2r^6}{50n^3r^6} =$

$\frac{20q^6b^6}{5q^3b^3} =$

$\frac{2q^6m^3}{10q^2m} =$

$\frac{2rh^2}{8r^4h} =$

$\frac{5n^5x^6}{20n^4x^4} =$

$\frac{6qp^5}{10q^6p^3} =$

$\frac{100n^4w^5}{10n^3w^4} =$

$\frac{25w^2t^3}{20w^6t^2} =$

$\frac{2f^2p^5}{4f^2p^2} =$

$\frac{9n^6a^6}{12n^4a^3} =$

$\frac{15k^4h}{3kh} =$

$\frac{10gf^5}{40g^6f} =$

$\frac{9n^3y^6}{6n^2y^2} =$

$\frac{2v^3g^2}{6v^6g} =$

$\frac{20ky^4}{16k^3y^6} =$

# Answer Key

Simplify each expression.

$$\frac{50u^5a}{40u^2a^5} = \frac{5u^3}{4a^4}$$

$$\frac{10k^5m^2}{6km^5} = \frac{5k^4}{3m^3}$$

$$\frac{9q^2g^6}{12q^4g^4} = \frac{3g^2}{4q^2}$$

$$\frac{20t^2m^5}{16t^2m^3} = \frac{5m^2}{4}$$

$$\frac{8w^6a^6}{4wa^2} = 2w^5a^4$$

$$\frac{5n^2r^6}{50n^3r^6} = \frac{1}{10n}$$

$$\frac{20q^6b^6}{5q^3b^3} = 4q^3b^3$$

$$\frac{2q^6m^3}{10q^2m} = \frac{q^4m^2}{5}$$

$$\frac{2rh^2}{8r^4h} = \frac{h}{4r^3}$$

$$\frac{5n^5x^6}{20n^4x^4} = \frac{nx^2}{4}$$

$$\frac{6qp^5}{10q^6p^3} = \frac{3p^2}{5q^5}$$

$$\frac{100n^4w^5}{10n^3w^4} = 10nw$$

$$\frac{25w^2t^3}{20w^6t^2} = \frac{5t}{4w^4}$$

$$\frac{2f^2p^5}{4f^2p^2} = \frac{p^3}{2}$$

$$\frac{9n^6a^6}{12n^4a^3} = \frac{3n^2a^3}{4}$$

$$\frac{15k^4h}{3kh} = 5k^3$$

$$\frac{10gf^5}{40g^6f} = \frac{f^4}{4g^5}$$

$$\frac{9n^3y^6}{6n^2y^2} = \frac{3ny^4}{2}$$

$$\frac{2v^3g^2}{6v^6g} = \frac{g}{3v^3}$$

$$\frac{20ky^4}{16k^3y^6} = \frac{5}{4k^2y^2}$$

Simplify each expression.

$\frac{3k^3a}{30k^5a^4} =$

$\frac{5b^2u}{50b^6u^2} =$

$\frac{8a^2x^4}{20a^6x^2} =$

$\frac{30a^3v^2}{3a^4v} =$

$\frac{20b^2k^3}{25b^3k^3} =$

$\frac{3a^6f^3}{30a^6f^2} =$

$\frac{6t^3f}{9tf^5} =$

$\frac{50p^4q^4}{40p^2q} =$

$\frac{9ry^6}{3r^6y^4} =$

$\frac{30k^4m^5}{3k^4m^4} =$

$\frac{16bk^6}{12b^3k^2} =$

$\frac{10r^4f^5}{15r^6f} =$

$\frac{40r^2u^3}{4ru^5} =$

$\frac{4t^2r^6}{6tr^2} =$

$\frac{12u^3n^4}{16u^3n^3} =$

$\frac{10d^3k}{100d^6k} =$

$\frac{8w^5v^5}{2w^5v^3} =$

$\frac{5d^2a^5}{10d^2a^5} =$

$\frac{100n^6a}{10n^4a^3} =$

$\frac{6f^5w^4}{3fw^2} =$

# Answer Key

Simplify each expression.

$$\frac{3k^3a}{30k^5a^4} = \frac{1}{10k^2a^3}$$

$$\frac{8a^2x^4}{20a^6x^2} = \frac{2x^2}{5a^4}$$

$$\frac{20b^2k^3}{25b^3k^3} = \frac{4}{5b}$$

$$\frac{6t^3f}{9tf^5} = \frac{2t^2}{3f^4}$$

$$\frac{9ry^6}{3r^6y^4} = \frac{3y^2}{r^5}$$

$$\frac{16bk^6}{12b^3k^2} = \frac{4k^4}{3b^2}$$

$$\frac{40r^2u^3}{4ru^5} = \frac{10r}{u^2}$$

$$\frac{12u^3n^4}{16u^3n^3} = \frac{3n}{4}$$

$$\frac{8w^5v^5}{2w^5v^3} = 4v^2$$

$$\frac{100n^6a}{10n^4a^3} = \frac{10n^2}{a^2}$$

$$\frac{5b^2u}{50b^6u^2} = \frac{1}{10b^4u}$$

$$\frac{30a^3v^2}{3a^4v} = \frac{10v}{a}$$

$$\frac{3a^6f^3}{30a^6f^2} = \frac{f}{10}$$

$$\frac{50p^4q^4}{40p^2q} = \frac{5p^2q^3}{4}$$

$$\frac{30k^4m^5}{3k^4m^4} = 10m$$

$$\frac{10r^4f^5}{15r^6f} = \frac{2f^4}{3r^2}$$

$$\frac{4t^2r^6}{6tr^2} = \frac{2tr^4}{3}$$

$$\frac{10d^3k}{100d^6k} = \frac{1}{10d^3}$$

$$\frac{5d^2a^5}{10d^2a^5} = \frac{1}{2}$$

$$\frac{6f^5w^4}{3fw^2} = 2f^4w^2$$

Simplify each expression.

$\frac{10fh^5}{15f^2h^4} =$

$\frac{12dn^2}{16d^5n^3} =$

$\frac{5b^4u}{25b^2u^2} =$

$\frac{100u^5q}{10uq^2} =$

$\frac{12yb^3}{16y^4b^2} =$

$\frac{30w^6h}{3w^3h^2} =$

$\frac{20r^4v^4}{50r^5v^2} =$

$\frac{8y^4d^5}{10y^5d^5} =$

$\frac{15m^4x^2}{20m^3x^3} =$

$\frac{15kt^3}{25k^6t^3} =$

$\frac{16a^3p^3}{12a^6p^6} =$

$\frac{16g^3v^2}{20g^6v^4} =$

$\frac{6by^5}{10b^6y^2} =$

$\frac{4hd^6}{10h^5d^6} =$

$\frac{6a^6w^2}{15a^5w} =$

$\frac{5g^5m^3}{10g^5m^4} =$

$\frac{20q^3a^3}{30q^3a^6} =$

$\frac{15rm^5}{5r^5m^3} =$

$\frac{2bn^3}{4b^2n^2} =$

$\frac{25b^2p}{5b^6p^4} =$

# Answer Key

Simplify each expression.

$$\frac{10fh^5}{15f^2h^4} = \frac{2h}{3f}$$

$$\frac{5b^4u}{25b^2u^2} = \frac{b^2}{5u}$$

$$\frac{12yb^3}{16y^4b^2} = \frac{3b}{4y^3}$$

$$\frac{20r^4v^4}{50r^5v^2} = \frac{2v^2}{5r}$$

$$\frac{15m^4x^2}{20m^3x^3} = \frac{3m}{4x}$$

$$\frac{16a^3p^3}{12a^6p^6} = \frac{4}{3a^3p^3}$$

$$\frac{6by^5}{10b^6y^2} = \frac{3y^3}{5b^5}$$

$$\frac{6a^6w^2}{15a^5w} = \frac{2aw}{5}$$

$$\frac{20q^3a^3}{30q^3a^6} = \frac{2}{3a^3}$$

$$\frac{2bn^3}{4b^2n^2} = \frac{n}{2b}$$

$$\frac{12dn^2}{16d^5n^3} = \frac{3}{4d^4n}$$

$$\frac{100u^5q}{10uq^2} = \frac{10u^4}{q}$$

$$\frac{30w^6h}{3w^3h^2} = \frac{10w^3}{h}$$

$$\frac{8y^4d^5}{10y^5d^5} = \frac{4}{5y}$$

$$\frac{15kt^3}{25k^6t^3} = \frac{3}{5k^5}$$

$$\frac{16g^3v^2}{20g^6v^4} = \frac{4}{5g^3v^2}$$

$$\frac{4hd^6}{10h^5d^6} = \frac{2}{5h^4}$$

$$\frac{5g^5m^3}{10g^5m^4} = \frac{1}{2m}$$

$$\frac{15rm^5}{5r^5m^3} = \frac{3m^2}{r^4}$$

$$\frac{25b^2p}{5b^6p^4} = \frac{5}{b^4p^3}$$

Simplify each expression.

$\frac{9mu}{6m^3u} =$

$\frac{3vf}{15v^5f^3} =$

$\frac{25w^5q^3}{15w^3q^3} =$

$\frac{30u^6g^4}{50u^4g^5} =$

$\frac{6f^2y^3}{15f^2y^3} =$

$\frac{15h^2b^3}{6h^2b} =$

$\frac{6g^5d^5}{3g^5d^6} =$

$\frac{16q^3u^3}{20q^2u^2} =$

$\frac{10d^2b^2}{5d^3b^4} =$

$\frac{30m^3k^2}{40m^5k^3} =$

$\frac{8a^5q^6}{4a^2q^6} =$

$\frac{20v^3d^3}{4v^5d^2} =$

$\frac{4f^2q^3}{10f^2q^4} =$

$\frac{4w^2y^2}{8w^2y} =$

$\frac{100v^6b^4}{10v^5b^5} =$

$\frac{15f^5b^5}{6f^2b^2} =$

$\frac{4n^3r^2}{6n^5r^4} =$

$\frac{30n^4a^3}{20na^2} =$

$\frac{3y^4k^2}{12y^6k^2} =$

$\frac{10d^3n^6}{50d^5n^5} =$

# Answer Key

Simplify each expression.

$$\frac{9mu}{6m^3u} = \frac{3}{2m^2}$$

$$\frac{3vf}{15v^5f^3} = \frac{1}{5v^4f^2}$$

$$\frac{25w^5q^3}{15w^3q^3} = \frac{5w^2}{3}$$

$$\frac{30u^6g^4}{50u^4g^5} = \frac{3u^2}{5g}$$

$$\frac{6f^2y^3}{15f^2y^3} = \frac{2}{5}$$

$$\frac{15h^2b^3}{6h^2b} = \frac{5b^2}{2}$$

$$\frac{6g^5d^5}{3g^5d^6} = \frac{2}{d}$$

$$\frac{16q^3u^3}{20q^2u^2} = \frac{4qu}{5}$$

$$\frac{10d^2b^2}{5d^3b^4} = \frac{2}{db^2}$$

$$\frac{30m^3k^2}{40m^5k^3} = \frac{3}{4m^2k}$$

$$\frac{8a^5q^6}{4a^2q^6} = 2a^3$$

$$\frac{20v^3d^3}{4v^5d^2} = \frac{5d}{v^2}$$

$$\frac{4f^2q^3}{10f^2q^4} = \frac{2}{5q}$$

$$\frac{4w^2y^2}{8w^2y} = \frac{y}{2}$$

$$\frac{100v^6b^4}{10v^5b^5} = \frac{10v}{b}$$

$$\frac{15f^5b^5}{6f^2b^2} = \frac{5f^3b^3}{2}$$

$$\frac{4n^3r^2}{6n^5r^4} = \frac{2}{3n^2r^2}$$

$$\frac{30n^4a^3}{20na^2} = \frac{3n^3a}{2}$$

$$\frac{3y^4k^2}{12y^6k^2} = \frac{1}{4y^2}$$

$$\frac{10d^3n^6}{50d^5n^5} = \frac{n}{5d^2}$$

Simplify each expression.

$$\frac{2n^5x^4}{20n^5x^6} =$$

$$\frac{5b^2p^3}{50b^6p^2} =$$

$$\frac{15f^3m^5}{6f^3m^3} =$$

$$\frac{5r^3b^3}{15r^5b^3} =$$

$$\frac{9g^5y}{3g^3y^3} =$$

$$\frac{20y^5g^2}{8yg^2} =$$

$$\frac{25a^4m^4}{5a^6m^6} =$$

$$\frac{40x^2k^3}{4xk^6} =$$

$$\frac{2nk^6}{20n^5k^6} =$$

$$\frac{20r^2d^4}{30r^6d^5} =$$

$$\frac{3w^4g}{30w^5g^2} =$$

$$\frac{2u^2b^3}{6u^6b^6} =$$

$$\frac{100w^2f}{10w^3f^2} =$$

$$\frac{2q^5x}{6q^4x^6} =$$

$$\frac{15t^6u^6}{6t^6u^3} =$$

$$\frac{50q^4h^6}{30qh} =$$

$$\frac{9y^6f^2}{12y^4f^4} =$$

$$\frac{10k^2r^6}{15k^5r^6} =$$

$$\frac{20a^6y^5}{4a^4y^5} =$$

$$\frac{30b^6p^5}{40b^3p^2} =$$

# Answer Key

Simplify each expression.

$$\frac{2n^5x^4}{20n^5x^6} = \frac{1}{10x^2}$$

$$\frac{3w^4g}{30w^5g^2} = \frac{1}{10wg}$$

$$\frac{5b^2p^3}{50b^6p^2} = \frac{p}{10b^4}$$

$$\frac{2u^2b^3}{6u^6b^6} = \frac{1}{3u^4b^3}$$

$$\frac{15f^3m^5}{6f^3m^3} = \frac{5m^2}{2}$$

$$\frac{100w^2f}{10w^3f^2} = \frac{10}{wf}$$

$$\frac{5r^3b^3}{15r^5b^3} = \frac{1}{3r^2}$$

$$\frac{2q^5x}{6q^4x^6} = \frac{q}{3x^5}$$

$$\frac{9g^5y}{3g^3y^3} = \frac{3g^2}{y^2}$$

$$\frac{15t^6u^6}{6t^6u^3} = \frac{5u^3}{2}$$

$$\frac{20y^5g^2}{8yg^2} = \frac{5y^4}{2}$$

$$\frac{50q^4h^6}{30qh} = \frac{5q^3h^5}{3}$$

$$\frac{25a^4m^4}{5a^6m^6} = \frac{5}{a^2m^2}$$

$$\frac{9y^6f^2}{12y^4f^4} = \frac{3y^2}{4f^2}$$

$$\frac{40x^2k^3}{4xk^6} = \frac{10x}{k^3}$$

$$\frac{10k^2r^6}{15k^5r^6} = \frac{2}{3k^3}$$

$$\frac{2nk^6}{20n^5k^6} = \frac{1}{10n^4}$$

$$\frac{20a^6y^5}{4a^4y^5} = 5a^2$$

$$\frac{20r^2d^4}{30r^6d^5} = \frac{2}{3r^4d}$$

$$\frac{30b^6p^5}{40b^3p^2} = \frac{3b^3p^3}{4}$$

Simplify each expression.

$\frac{2f^4q^2}{10f^6q^4} =$

$\frac{15f^3r^2}{9f^5r^5} =$

$\frac{6mw^5}{10m^6w} =$

$\frac{8m^2d^3}{4m^2d^6} =$

$\frac{3x^2k}{15x^4k^3} =$

$\frac{20n^6a^4}{16n^3a} =$

$\frac{5p^6n^6}{10p^4n^3} =$

$\frac{15v^4f^3}{6v^2f^5} =$

$\frac{3b^3x^6}{6b^3x^6} =$

$\frac{10k^2q^6}{40k^5q} =$

$\frac{50k^2f^5}{20k^3f^4} =$

$\frac{16g^3n^5}{4gn} =$

$\frac{10f^2d^5}{40f^2d} =$

$\frac{30g^4f^5}{3gf^4} =$

$\frac{40d^3y^3}{4d^3y^6} =$

$\frac{50r^4v}{40r^4v} =$

$\frac{12v^6t^3}{4v^6t^4} =$

$\frac{2k^6w^5}{20k^4w^6} =$

$\frac{30fd}{20f^2d^5} =$

$\frac{15t^3p^4}{25t^6p^2} =$

# Answer Key

Simplify each expression.

$$\frac{2f^4q^2}{10f^6q^4} = \frac{1}{5f^2q^2}$$

$$\frac{15f^3r^2}{9f^5r^5} = \frac{5}{3f^2r^3}$$

$$\frac{6mw^5}{10m^6w} = \frac{3w^4}{5m^5}$$

$$\frac{8m^2d^3}{4m^2d^6} = \frac{2}{d^3}$$

$$\frac{3x^2k}{15x^4k^3} = \frac{1}{5x^2k^2}$$

$$\frac{20n^6a^4}{16n^3a} = \frac{5n^3a^3}{4}$$

$$\frac{5p^6n^6}{10p^4n^3} = \frac{p^2n^3}{2}$$

$$\frac{15v^4f^3}{6v^2f^5} = \frac{5v^2}{2f^2}$$

$$\frac{3b^3x^6}{6b^3x^6} = \frac{1}{2}$$

$$\frac{10k^2q^6}{40k^5q} = \frac{q^5}{4k^3}$$

$$\frac{50k^2f^5}{20k^3f^4} = \frac{5f}{2k}$$

$$\frac{16g^3n^5}{4gn} = 4g^2n^4$$

$$\frac{10f^2d^5}{40f^2d} = \frac{d^4}{4}$$

$$\frac{30g^4f^5}{3gf^4} = 10g^3f$$

$$\frac{40d^3y^3}{4d^3y^6} = \frac{10}{y^3}$$

$$\frac{50r^4v}{40r^4v} = \frac{5}{4}$$

$$\frac{12v^6t^3}{4v^6t^4} = \frac{3}{t}$$

$$\frac{2k^6w^5}{20k^4w^6} = \frac{k^2}{10w}$$

$$\frac{30fd}{20f^2d^5} = \frac{3}{2fd^4}$$

$$\frac{15t^3p^4}{25t^6p^2} = \frac{3p^2}{5t^3}$$

Simplify each expression.

$\frac{20b^2d^6}{5bd^5} =$

$\frac{50g^2u^2}{5g^5u^5} =$

$\frac{3w^3d^2}{6w^3d} =$

$\frac{2v^3t^4}{10v^5t} =$

$\frac{20h^4q^6}{12h^3q^5} =$

$\frac{9r^5n^3}{12r^6n^5} =$

$\frac{4d^4f^4}{16df} =$

$\frac{6p^4h^4}{9ph^6} =$

$\frac{9d^4x}{3dx^4} =$

$\frac{8g^6b^4}{10g^4b^3} =$

$\frac{20a^2f^2}{5a^2f^4} =$

$\frac{2g^3w}{4g^6w^5} =$

$\frac{20a^6w}{4a^4w} =$

$\frac{40y^2v^3}{10y^3v^4} =$

$\frac{10y^6t^6}{20y^4t^6} =$

$\frac{10by^4}{40b^6y^6} =$

$\frac{30t^2w^4}{10t^3w} =$

$\frac{12w^6p^2}{16w^3p^3} =$

$\frac{15f^4r^5}{20fr^3} =$

$\frac{10d^4f^2}{4d^2f} =$

# Answer Key

Simplify each expression.

$$\frac{20b^2d^6}{5bd^5} = 4bd$$

$$\frac{3w^3d^2}{6w^3d} = \frac{d}{2}$$

$$\frac{20h^4q^6}{12h^3q^5} = \frac{5hq}{3}$$

$$\frac{4d^4f^4}{16df} = \frac{d^3f^3}{4}$$

$$\frac{9d^4x}{3dx^4} = \frac{3d^3}{x^3}$$

$$\frac{20a^2f^2}{5a^2f^4} = \frac{4}{f^2}$$

$$\frac{20a^6w}{4a^4w} = 5a^2$$

$$\frac{10y^6t^6}{20y^4t^6} = \frac{y^2}{2}$$

$$\frac{30t^2w^4}{10t^3w} = \frac{3w^3}{t}$$

$$\frac{15f^4r^5}{20fr^3} = \frac{3f^3r^2}{4}$$

$$\frac{50g^2u^2}{5g^5u^5} = \frac{10}{g^3u^3}$$

$$\frac{2v^3t^4}{10v^5t} = \frac{t^3}{5v^2}$$

$$\frac{9r^5n^3}{12r^6n^5} = \frac{3}{4rn^2}$$

$$\frac{6p^4h^4}{9ph^6} = \frac{2p^3}{3h^2}$$

$$\frac{8g^6b^4}{10g^4b^3} = \frac{4g^2b}{5}$$

$$\frac{2g^3w}{4g^6w^5} = \frac{1}{2g^3w^4}$$

$$\frac{40y^2v^3}{10y^3v^4} = \frac{4}{yv}$$

$$\frac{10by^4}{40b^6y^6} = \frac{1}{4b^5y^2}$$

$$\frac{12w^6p^2}{16w^3p^3} = \frac{3w^3}{4p}$$

$$\frac{10d^4f^2}{4d^2f} = \frac{5d^2f}{2}$$

Simplify each expression.

$\frac{5q^2d}{10q^6d^3} =$

$\frac{20gh^6}{5g^4h} =$

$\frac{6v^5x}{4vx^2} =$

$\frac{30x^6k^6}{50x^4k^2} =$

$\frac{16vh}{4v^5h^6} =$

$\frac{10t^3k^2}{15tk^2} =$

$\frac{10f^6b^6}{100f^3b^3} =$

$\frac{20f^3p^3}{50fp^3} =$

$\frac{30d^6r^4}{20d^2r^3} =$

$\frac{15h^3x}{20h^4x^2} =$

$\frac{25w^2g^4}{20w^2g} =$

$\frac{50na}{20n^2a^4} =$

$\frac{20n^5q^5}{2n^2q^6} =$

$\frac{16m^4h^5}{20m^5h^3} =$

$\frac{8w^6m^4}{12w^3m^3} =$

$\frac{6u^6r^6}{9u^2r^5} =$

$\frac{50n^6q}{10n^3q^5} =$

$\frac{15qv^4}{5q^2v} =$

$\frac{4n^3k^2}{16n^2k^6} =$

$\frac{10m^2d^4}{30md^2} =$

# Answer Key

Simplify each expression.

$$\frac{5q^2d}{10q^6d^3} = \frac{1}{2q^4d^2}$$

$$\frac{20gh^6}{5g^4h} = \frac{4h^5}{g^3}$$

$$\frac{6v^5x}{4vx^2} = \frac{3v^4}{2x}$$

$$\frac{30x^6k^6}{50x^4k^2} = \frac{3x^2k^4}{5}$$

$$\frac{16vh}{4v^5h^6} = \frac{4}{v^4h^5}$$

$$\frac{10t^3k^2}{15tk^2} = \frac{2t^2}{3}$$

$$\frac{10f^6b^6}{100f^3b^3} = \frac{f^3b^3}{10}$$

$$\frac{20f^3p^3}{50fp^3} = \frac{2f^2}{5}$$

$$\frac{30d^6r^4}{20d^2r^3} = \frac{3d^4r}{2}$$

$$\frac{15h^3x}{20h^4x^2} = \frac{3}{4hx}$$

$$\frac{25w^2g^4}{20w^2g} = \frac{5g^3}{4}$$

$$\frac{50na}{20n^2a^4} = \frac{5}{2na^3}$$

$$\frac{20n^5q^5}{2n^2q^6} = \frac{10n^3}{q}$$

$$\frac{16m^4h^5}{20m^5h^3} = \frac{4h^2}{5m}$$

$$\frac{8w^6m^4}{12w^3m^3} = \frac{2w^3m}{3}$$

$$\frac{6u^6r^6}{9u^2r^5} = \frac{2u^4r}{3}$$

$$\frac{50n^6q}{10n^3q^5} = \frac{5n^3}{q^4}$$

$$\frac{15qv^4}{5q^2v} = \frac{3v^3}{q}$$

$$\frac{4n^3k^2}{16n^2k^6} = \frac{n}{4k^4}$$

$$\frac{10m^2d^4}{30md^2} = \frac{md^2}{3}$$

Simplify each expression.

$\frac{20h^6y^2}{4h^2y^2} =$

$\frac{20w^5v^6}{25w^4v^6} =$

$\frac{15b^6g^4}{20b^4g^4} =$

$\frac{20t^4b^5}{10t^4b^2} =$

$\frac{15t^2n^6}{3t^5n^5} =$

$\frac{12x^4g^4}{9x^3g^3} =$

$\frac{3m^2f^5}{9m^5f^5} =$

$\frac{30hd^5}{10h^3d^3} =$

$\frac{25f^4h^3}{20f^6h^3} =$

$\frac{50w^6q}{10w^2q^2} =$

$\frac{5q^5w}{15q^4w^2} =$

$\frac{15n^4m^4}{25n^4m^2} =$

$\frac{12b^4y^2}{9b^3y} =$

$\frac{25hf^4}{15h^6f} =$

$\frac{8dm^2}{4d^2m} =$

$\frac{2v^4d^5}{6v^3d^4} =$

$\frac{6r^2q^3}{8r^4q^5} =$

$\frac{25kv^6}{5k^4v} =$

$\frac{8ug^6}{12u^2g} =$

$\frac{100hg^2}{10h^2g^2} =$

# Answer Key

Simplify each expression.

$$\frac{20h^6y^2}{4h^2y^2} = 5h^4$$

$$\frac{20w^5v^6}{25w^4v^6} = \frac{4w}{5}$$

$$\frac{15b^6g^4}{20b^4g^4} = \frac{3b^2}{4}$$

$$\frac{20t^4b^5}{10t^4b^2} = 2b^3$$

$$\frac{15t^2n^6}{3t^5n^5} = \frac{5n}{t^3}$$

$$\frac{12x^4g^4}{9x^3g^3} = \frac{4xg}{3}$$

$$\frac{3m^2f^5}{9m^5f^5} = \frac{1}{3m^3}$$

$$\frac{30hd^5}{10h^3d^3} = \frac{3d^2}{h^2}$$

$$\frac{25f^4h^3}{20f^6h^3} = \frac{5}{4f^2}$$

$$\frac{50w^6q}{10w^2q^2} = \frac{5w^4}{q}$$

$$\frac{5q^5w}{15q^4w^2} = \frac{q}{3w}$$

$$\frac{15n^4m^4}{25n^4m^2} = \frac{3m^2}{5}$$

$$\frac{12b^4y^2}{9b^3y} = \frac{4by}{3}$$

$$\frac{25hf^4}{15h^6f} = \frac{5f^3}{3h^5}$$

$$\frac{8dm^2}{4d^2m} = \frac{2m}{d}$$

$$\frac{2v^4d^5}{6v^3d^4} = \frac{vd}{3}$$

$$\frac{6r^2q^3}{8r^4q^5} = \frac{3}{4r^2q^2}$$

$$\frac{25kv^6}{5k^4v} = \frac{5v^5}{k^3}$$

$$\frac{8ug^6}{12u^2g} = \frac{2g^5}{3u}$$

$$\frac{100hg^2}{10h^2g^2} = \frac{10}{h}$$

Simplify each expression.

$$\frac{40n^4f^2}{30n^2f^5} =$$

$$\frac{3b^5h^3}{30b^2h^2} =$$

$$\frac{10qx^4}{25q^2x^6} =$$

$$\frac{5u^3x^5}{15ux^2} =$$

$$\frac{15h^6t^5}{9h^3t^3} =$$

$$\frac{20g^5w^3}{30g^4w^6} =$$

$$\frac{30y^2q^6}{20y^6q^5} =$$

$$\frac{12t^6h^3}{15th^2} =$$

$$\frac{50w^5q}{10w^2q^2} =$$

$$\frac{12w^2u}{20w^6u^4} =$$

$$\frac{3b^2w^3}{30b^3w^2} =$$

$$\frac{2ty}{6t^6y} =$$

$$\frac{20k^2t}{10k^4t^2} =$$

$$\frac{16g^5w}{4g^3w} =$$

$$\frac{10n^2b^3}{6n^4b^3} =$$

$$\frac{4y^6p}{20yp^5} =$$

$$\frac{2t^3x}{4tx^2} =$$

$$\frac{40u^5h^2}{50u^3h^5} =$$

$$\frac{6b^3f^4}{9b^6f^6} =$$

$$\frac{2u^2w^5}{8u^3w^6} =$$

# Answer Key

Simplify each expression.

$$\frac{40n^4f^2}{30n^2f^5} = \frac{4n^2}{3f^3}$$

$$\frac{3b^2w^3}{30b^3w^2} = \frac{w}{10b}$$

$$\frac{3b^5h^3}{30b^2h^2} = \frac{b^3h}{10}$$

$$\frac{2ty}{6t^6y} = \frac{1}{3t^5}$$

$$\frac{10qx^4}{25q^2x^6} = \frac{2}{5qx^2}$$

$$\frac{20k^2t}{10k^4t^2} = \frac{2}{k^2t}$$

$$\frac{5u^3x^5}{15ux^2} = \frac{u^2x^3}{3}$$

$$\frac{16g^5w}{4g^3w} = 4g^2$$

$$\frac{15h^6t^5}{9h^3t^3} = \frac{5h^3t^2}{3}$$

$$\frac{10n^2b^3}{6n^4b^3} = \frac{5}{3n^2}$$

$$\frac{20g^5w^3}{30g^4w^6} = \frac{2g}{3w^3}$$

$$\frac{4y^6p}{20yp^5} = \frac{y^5}{5p^4}$$

$$\frac{30y^2q^6}{20y^6q^5} = \frac{3q}{2y^4}$$

$$\frac{2t^3x}{4tx^2} = \frac{t^2}{2x}$$

$$\frac{12t^6h^3}{15th^2} = \frac{4t^5h}{5}$$

$$\frac{40u^5h^2}{50u^3h^5} = \frac{4u^2}{5h^3}$$

$$\frac{50w^5q}{10w^2q^2} = \frac{5w^3}{q}$$

$$\frac{6b^3f^4}{9b^6f^6} = \frac{2}{3b^3f^2}$$

$$\frac{12w^2u}{20w^6u^4} = \frac{3}{5w^4u^3}$$

$$\frac{2u^2w^5}{8u^3w^6} = \frac{1}{4uw}$$

Simplify each expression.

$\frac{15k^2g^3}{9k^4g^2} =$

$\frac{2w^4b^5}{20w^5b^2} =$

$\frac{3q^4p}{15q^6p^6} =$

$\frac{25g^4q^3}{10g^2q^2} =$

$\frac{12p^2x^2}{20p^6x^2} =$

$\frac{8aq^5}{4aq^3} =$

$\frac{10at}{8a^6t^2} =$

$\frac{15dy}{25d^5y^5} =$

$\frac{4m^4v^2}{10m^6v^4} =$

$\frac{10hg^3}{6h^2g^5} =$

$\frac{3r^5q^6}{12rq} =$

$\frac{9m^6g}{15mg} =$

$\frac{20t^6v^3}{5t^2v^6} =$

$\frac{6x^3b}{10x^6b^6} =$

$\frac{25p^2u^2}{15p^4u} =$

$\frac{20b^3h^2}{8bh^2} =$

$\frac{25m^2v}{10m^5v^6} =$

$\frac{10f^6u^4}{4fu} =$

$\frac{2tg^6}{20t^5g^5} =$

$\frac{20g^2f^5}{16g^5f^3} =$

# Answer Key

Simplify each expression.

$$\frac{15k^2g^3}{9k^4g^2} = \frac{5g}{3k^2}$$

$$\frac{3q^4p}{15q^6p^6} = \frac{1}{5q^2p^5}$$

$$\frac{12p^2x^2}{20p^6x^2} = \frac{3}{5p^4}$$

$$\frac{10at}{8a^6t^2} = \frac{5}{4a^5t}$$

$$\frac{4m^4v^2}{10m^6v^4} = \frac{2}{5m^2v^2}$$

$$\frac{3r^5q^6}{12rq} = \frac{r^4q^5}{4}$$

$$\frac{20t^6v^3}{5t^2v^6} = \frac{4t^4}{v^3}$$

$$\frac{25p^2u^2}{15p^4u} = \frac{5u}{3p^2}$$

$$\frac{25m^2v}{10m^5v^6} = \frac{5}{2m^3v^5}$$

$$\frac{2tg^6}{20t^5g^5} = \frac{g}{10t^4}$$

$$\frac{2w^4b^5}{20w^5b^2} = \frac{b^3}{10w}$$

$$\frac{25g^4q^3}{10g^2q^2} = \frac{5g^2q}{2}$$

$$\frac{8aq^5}{4aq^3} = 2q^2$$

$$\frac{15dy}{25d^5y^5} = \frac{3}{5d^4y^4}$$

$$\frac{10hg^3}{6h^2g^5} = \frac{5}{3hg^2}$$

$$\frac{9m^6g}{15mg} = \frac{3m^5}{5}$$

$$\frac{6x^3b}{10x^6b^6} = \frac{3}{5x^3b^5}$$

$$\frac{20b^3h^2}{8bh^2} = \frac{5b^2}{2}$$

$$\frac{10f^6u^4}{4fu} = \frac{5f^5u^3}{2}$$

$$\frac{20g^2f^5}{16g^5f^3} = \frac{5f^2}{4g^3}$$

Simplify each expression.

$\frac{2x^4a^4}{6x^5a^2} =$

$\frac{50a^2m^5}{30a^6m^3} =$

$\frac{10f^5t^3}{25ft^2} =$

$\frac{20t^3b^4}{10t^2b^4} =$

$\frac{10a^3p^4}{25a^3p^2} =$

$\frac{12yq^4}{16y^6q^4} =$

$\frac{20t^6k^2}{30t^2k^3} =$

$\frac{15y^2q^4}{6y^5q^4} =$

$\frac{10b^5n^5}{25b^4n} =$

$\frac{40p^2n^5}{4pn^3} =$

$\frac{20g^4q^5}{5g^6q^4} =$

$\frac{3w^3q^5}{6w^3q^2} =$

$\frac{10a^2v^5}{20a^3v^5} =$

$\frac{10g^5k^5}{6gk^3} =$

$\frac{40n^2h^6}{10n^4h^2} =$

$\frac{40t^3b^5}{30t^4b^4} =$

$\frac{15gt^2}{20gt} =$

$\frac{15t^5u^2}{9t^2u} =$

$\frac{20t^5k^4}{4tk^3} =$

$\frac{20b^2t^4}{30b^3t} =$

# Answer Key

Simplify each expression.

$$\frac{2x^4a^4}{6x^5a^2} = \frac{a^2}{3x}$$

$$\frac{10f^5t^3}{25ft^2} = \frac{2f^4t}{5}$$

$$\frac{10a^3p^4}{25a^3p^2} = \frac{2p^2}{5}$$

$$\frac{20t^6k^2}{30t^2k^3} = \frac{2t^4}{3k}$$

$$\frac{10b^5n^5}{25b^4n} = \frac{2bn^4}{5}$$

$$\frac{20g^4q^5}{5g^6q^4} = \frac{4q}{g^2}$$

$$\frac{10a^2v^5}{20a^3v^5} = \frac{1}{2a}$$

$$\frac{40n^2h^6}{10n^4h^2} = \frac{4h^4}{n^2}$$

$$\frac{15gt^2}{20gt} = \frac{3t}{4}$$

$$\frac{20t^5k^4}{4tk^3} = 5t^4k$$

$$\frac{50a^2m^5}{30a^6m^3} = \frac{5m^2}{3a^4}$$

$$\frac{20t^3b^4}{10t^2b^4} = 2t$$

$$\frac{12yq^4}{16y^6q^4} = \frac{3}{4y^5}$$

$$\frac{15y^2q^4}{6y^5q^4} = \frac{5}{2y^3}$$

$$\frac{40p^2n^5}{4pn^3} = 10pn^2$$

$$\frac{3w^3q^5}{6w^3q^2} = \frac{q^3}{2}$$

$$\frac{10g^5k^5}{6gk^3} = \frac{5g^4k^2}{3}$$

$$\frac{40t^3b^5}{30t^4b^4} = \frac{4b}{3t}$$

$$\frac{15t^5u^2}{9t^2u} = \frac{5t^3u}{3}$$

$$\frac{20b^2t^4}{30b^3t} = \frac{2t^3}{3b}$$

Simplify each expression.

$$\frac{3a^2x}{12a^2x^5} =$$

$$\frac{30a^2n^6}{50a^5n^5} =$$

$$\frac{100vu^2}{10v^2u^6} =$$

$$\frac{20v^5t^4}{16v^2t^2} =$$

$$\frac{20uw^5}{50uw^3} =$$

$$\frac{10bq^2}{2b^4q^4} =$$

$$\frac{30tx^3}{40t^2x^5} =$$

$$\frac{50n^2b^3}{20n^6b^6} =$$

$$\frac{12k^4b^5}{3k^6b} =$$

$$\frac{2v^6u}{20v^5u^2} =$$

$$\frac{20u^5b^3}{50u^5b^3} =$$

$$\frac{10q^5x^3}{50q^4x^3} =$$

$$\frac{8h^2b}{20hb^3} =$$

$$\frac{20h^4q}{16h^4q^3} =$$

$$\frac{25h^6u^3}{20h^4u^3} =$$

$$\frac{9m^5u^2}{12m^3u^2} =$$

$$\frac{5kr}{50k^5r^6} =$$

$$\frac{25v^4q^6}{5v^5q^2} =$$

$$\frac{16a^5m}{20a^2m} =$$

$$\frac{4p^4w}{12p^4w^5} =$$

# Answer Key

Simplify each expression.

$$\frac{3a^2x}{12a^2x^5} = \frac{1}{4x^4}$$

$$\frac{100vu^2}{10v^2u^6} = \frac{10}{vu^4}$$

$$\frac{20uw^5}{50uw^3} = \frac{2w^2}{5}$$

$$\frac{30tx^3}{40t^2x^5} = \frac{3}{4tx^2}$$

$$\frac{12k^4b^5}{3k^6b} = \frac{4b^4}{k^2}$$

$$\frac{20u^5b^3}{50u^5b^3} = \frac{2}{5}$$

$$\frac{8h^2b}{20hb^3} = \frac{2h}{5b^2}$$

$$\frac{25h^6u^3}{20h^4u^3} = \frac{5h^2}{4}$$

$$\frac{5kr}{50k^5r^6} = \frac{1}{10k^4r^5}$$

$$\frac{16a^5m}{20a^2m} = \frac{4a^3}{5}$$

$$\frac{30a^2n^6}{50a^5n^5} = \frac{3n}{5a^3}$$

$$\frac{20v^5t^4}{16v^2t^2} = \frac{5v^3t^2}{4}$$

$$\frac{10bq^2}{2b^4q^4} = \frac{5}{b^3q^2}$$

$$\frac{50n^2b^3}{20n^6b^6} = \frac{5}{2n^4b^3}$$

$$\frac{2v^6u}{20v^5u^2} = \frac{v}{10u}$$

$$\frac{10q^5x^3}{50q^4x^3} = \frac{q}{5}$$

$$\frac{20h^4q}{16h^4q^3} = \frac{5}{4q^2}$$

$$\frac{9m^5u^2}{12m^3u^2} = \frac{3m^2}{4}$$

$$\frac{25v^4q^6}{5v^5q^2} = \frac{5q^4}{v}$$

$$\frac{4p^4w}{12p^4w^5} = \frac{1}{3w^4}$$

Simplify each expression.

$\frac{2q^5d^2}{6q^5d^4} =$

$\frac{4h^2u^5}{6h^5u^2} =$

$\frac{10qy^5}{20q^2y^6} =$

$\frac{10g^5r^6}{15g^4r^2} =$

$\frac{30d^2f^3}{3d^2f^2} =$

$\frac{10m^3g^6}{50m^3g} =$

$\frac{10pq^3}{6p^2q} =$

$\frac{3t^2n^4}{12t^6n^3} =$

$\frac{15b^5v^2}{5b^6v^3} =$

$\frac{3h^3d}{12h^2d^2} =$

$\frac{15k^3h^2}{12k^3h^3} =$

$\frac{8x^2g^6}{4x^5g^4} =$

$\frac{20d^2n^2}{25d^5n^2} =$

$\frac{4q^4a^2}{8q^6a^2} =$

$\frac{20m^6w^2}{10mw^3} =$

$\frac{3y^3b^6}{6y^5b^6} =$

$\frac{15g^6v^3}{20g^3v^3} =$

$\frac{4rx^2}{8r^2x^6} =$

$\frac{4r^2u^3}{2r^3u^2} =$

$\frac{10d^2f^5}{40d^5f^3} =$

# Answer Key

Simplify each expression.

$$\frac{2q^5d^2}{6q^5d^4} = \frac{1}{3d^2}$$

$$\frac{4h^2u^5}{6h^5u^2} = \frac{2u^3}{3h^3}$$

$$\frac{10qy^5}{20q^2y^6} = \frac{1}{2qy}$$

$$\frac{10g^5r^6}{15g^4r^2} = \frac{2gr^4}{3}$$

$$\frac{30d^2f^3}{3d^2f^2} = 10f$$

$$\frac{10m^3g^6}{50m^3g} = \frac{g^5}{5}$$

$$\frac{10pq^3}{6p^2q} = \frac{5q^2}{3p}$$

$$\frac{3t^2n^4}{12t^6n^3} = \frac{n}{4t^4}$$

$$\frac{15b^5v^2}{5b^6v^3} = \frac{3}{bv}$$

$$\frac{3h^3d}{12h^2d^2} = \frac{h}{4d}$$

$$\frac{15k^3h^2}{12k^3h^3} = \frac{5}{4h}$$

$$\frac{8x^2g^6}{4x^5g^4} = \frac{2g^2}{x^3}$$

$$\frac{20d^2n^2}{25d^5n^2} = \frac{4}{5d^3}$$

$$\frac{4q^4a^2}{8q^6a^2} = \frac{1}{2q^2}$$

$$\frac{20m^6w^2}{10mw^3} = \frac{2m^5}{w}$$

$$\frac{3y^3b^6}{6y^5b^6} = \frac{1}{2y^2}$$

$$\frac{15g^6v^3}{20g^3v^3} = \frac{3g^3}{4}$$

$$\frac{4rx^2}{8r^2x^6} = \frac{1}{2rx^4}$$

$$\frac{4r^2u^3}{2r^3u^2} = \frac{2u}{r}$$

$$\frac{10d^2f^5}{40d^5f^3} = \frac{f^2}{4d^3}$$

Simplify each expression.

$$\frac{8n^2d^3}{10nd^6} =$$

$$\frac{9p^3d^3}{15p^3d^5} =$$

$$\frac{16h^2a}{20ha^3} =$$

$$\frac{4w^6g^3}{10w^3g^4} =$$

$$\frac{9p^2r^4}{3p^2r^3} =$$

$$\frac{10k^5n^2}{40kn} =$$

$$\frac{2y^5u^3}{8y^5u^6} =$$

$$\frac{8w^4y}{2w^4y^2} =$$

$$\frac{10p^2q^3}{100p^6q^3} =$$

$$\frac{30p^2a^6}{50p^4a} =$$

$$\frac{6h^4x^5}{9h^3x} =$$

$$\frac{10a^5n^4}{40a^2n^2} =$$

$$\frac{12d^2k^5}{3dk^4} =$$

$$\frac{4q^4p^2}{2q^3p^6} =$$

$$\frac{20m^5v^4}{10m^3v^2} =$$

$$\frac{10b^4u^3}{15bu^3} =$$

$$\frac{20kd^6}{4k^4d^2} =$$

$$\frac{3d^5u^6}{12d^3u^4} =$$

$$\frac{10p^2a^3}{15p^5a^5} =$$

$$\frac{10g^3t^3}{30g^3t^3} =$$

# Answer Key

Simplify each expression.

$$\frac{8n^2d^3}{10nd^6} = \frac{4n}{5d^3}$$

$$\frac{9p^3d^3}{15p^3d^5} = \frac{3}{5d^2}$$

$$\frac{16h^2a}{20ha^3} = \frac{4h}{5a^2}$$

$$\frac{4w^6g^3}{10w^3g^4} = \frac{2w^3}{5g}$$

$$\frac{9p^2r^4}{3p^2r^3} = 3r$$

$$\frac{10k^5n^2}{40kn} = \frac{k^4n}{4}$$

$$\frac{2y^5u^3}{8y^5u^6} = \frac{1}{4u^3}$$

$$\frac{8w^4y}{2w^4y^2} = \frac{4}{y}$$

$$\frac{10p^2q^3}{100p^6q^3} = \frac{1}{10p^4}$$

$$\frac{30p^2a^6}{50p^4a} = \frac{3a^5}{5p^2}$$

$$\frac{6h^4x^5}{9h^3x} = \frac{2hx^4}{3}$$

$$\frac{10a^5n^4}{40a^2n^2} = \frac{a^3n^2}{4}$$

$$\frac{12d^2k^5}{3dk^4} = 4dk$$

$$\frac{4q^4p^2}{2q^3p^6} = \frac{2q}{p^4}$$

$$\frac{20m^5v^4}{10m^3v^2} = 2m^2v^2$$

$$\frac{10b^4u^3}{15bu^3} = \frac{2b^3}{3}$$

$$\frac{20kd^6}{4k^4d^2} = \frac{5d^4}{k^3}$$

$$\frac{3d^5u^6}{12d^3u^4} = \frac{d^2u^2}{4}$$

$$\frac{10p^2a^3}{15p^5a^5} = \frac{2}{3p^3a^2}$$

$$\frac{10g^3t^3}{30g^3t^3} = \frac{1}{3}$$

Simplify each expression.

$\frac{10b^4n}{20bn^2} =$

$\frac{15d^2m^2}{20d^3m^4} =$

$\frac{50k^2f^2}{40kf^3} =$

$\frac{2t^3h^6}{4t^3h} =$

$\frac{5t^6h^3}{25th^5} =$

$\frac{20h^6w^3}{30h^2w^2} =$

$\frac{20p^5y^3}{4p^6y^4} =$

$\frac{10b^4d^2}{4bd^5} =$

$\frac{4wp}{10w^5p^4} =$

$\frac{30b^6g^6}{10b^4g^6} =$

$\frac{4f^4b^2}{16f^4b^2} =$

$\frac{10ur^2}{15ur^2} =$

$\frac{2n^4w^5}{4n^6w^5} =$

$\frac{10k^6g^2}{5k^2g^3} =$

$\frac{25y^4v^3}{5y^2v} =$

$\frac{6a^4v^6}{9av^3} =$

$\frac{10p^4m^6}{8p^6m^2} =$

$\frac{20a^3r}{8a^4r^2} =$

$\frac{3t^4x^6}{15t^2x^3} =$

$\frac{8f^2p^6}{4f^5p^5} =$

# Answer Key

Simplify each expression.

$$\frac{10b^4n}{20bn^2} = \frac{b^3}{2n}$$

$$\frac{15d^2m^2}{20d^3m^4} = \frac{3}{4dm^2}$$

$$\frac{50k^2f^2}{40kf^3} = \frac{5k}{4f}$$

$$\frac{2t^3h^6}{4t^3h} = \frac{h^5}{2}$$

$$\frac{5t^6h^3}{25th^5} = \frac{t^5}{5h^2}$$

$$\frac{20h^6w^3}{30h^2w^2} = \frac{2h^4w}{3}$$

$$\frac{20p^5y^3}{4p^6y^4} = \frac{5}{py}$$

$$\frac{10b^4d^2}{4bd^5} = \frac{5b^3}{2d^3}$$

$$\frac{4wp}{10w^5p^4} = \frac{2}{5w^4p^3}$$

$$\frac{30b^6g^6}{10b^4g^6} = 3b^2$$

$$\frac{4f^4b^2}{16f^4b^2} = \frac{1}{4}$$

$$\frac{10ur^2}{15ur^2} = \frac{2}{3}$$

$$\frac{2n^4w^5}{4n^6w^5} = \frac{1}{2n^2}$$

$$\frac{10k^6g^2}{5k^2g^3} = \frac{2k^4}{g}$$

$$\frac{25y^4v^3}{5y^2v} = 5y^2v^2$$

$$\frac{6a^4v^6}{9av^3} = \frac{2a^3v^3}{3}$$

$$\frac{10p^4m^6}{8p^6m^2} = \frac{5m^4}{4p^2}$$

$$\frac{20a^3r}{8a^4r^2} = \frac{5}{2ar}$$

$$\frac{3t^4x^6}{15t^2x^3} = \frac{t^2x^3}{5}$$

$$\frac{8f^2p^6}{4f^5p^5} = \frac{2p}{f^3}$$

Simplify each expression.

$\frac{12p^2d}{9pd^4} =$

$\frac{3a^4d^4}{15a^4d^6} =$

$\frac{2p^2k^2}{4p^3k} =$

$\frac{30p^2x^4}{20px} =$

$\frac{8u^5y^2}{10u^4y^4} =$

$\frac{4gr^2}{12g^6r^2} =$

$\frac{50y^2f^4}{40y^3f^6} =$

$\frac{6q^2k^2}{3q^3k^3} =$

$\frac{20h^2k^3}{25h^3k^2} =$

$\frac{30r^2t^4}{40r^5t^4} =$

$\frac{40h^2q^2}{4hq^2} =$

$\frac{20t^4d^5}{5t^4d^4} =$

$\frac{6g^3t^2}{3g^4t^2} =$

$\frac{9k^3b^4}{12k^5b^2} =$

$\frac{30n^2x}{50n^3x^5} =$

$\frac{12t^3a^5}{20t^4a^3} =$

$\frac{3n^2x^5}{9n^4x^6} =$

$\frac{8m^2q^3}{12m^3q^5} =$

$\frac{16w^2y^2}{4w^3y^2} =$

$\frac{15t^6w^2}{25t^3w^2} =$

# Answer Key

Simplify each expression.

$$\frac{12p^2d}{9pd^4} = \frac{4p}{3d^3}$$

$$\frac{3a^4d^4}{15a^4d^6} = \frac{1}{5d^2}$$

$$\frac{2p^2k^2}{4p^3k} = \frac{k}{2p}$$

$$\frac{30p^2x^4}{20px} = \frac{3px^3}{2}$$

$$\frac{8u^5y^2}{10u^4y^4} = \frac{4u}{5y^2}$$

$$\frac{4gr^2}{12g^6r^2} = \frac{1}{3g^5}$$

$$\frac{50y^2f^4}{40y^3f^6} = \frac{5}{4yf^2}$$

$$\frac{6q^2k^2}{3q^3k^3} = \frac{2}{qk}$$

$$\frac{20h^2k^3}{25h^3k^2} = \frac{4k}{5h}$$

$$\frac{30r^2t^4}{40r^5t^4} = \frac{3}{4r^3}$$

$$\frac{40h^2q^2}{4hq^2} = 10h$$

$$\frac{20t^4d^5}{5t^4d^4} = 4d$$

$$\frac{6g^3t^2}{3g^4t^2} = \frac{2}{g}$$

$$\frac{9k^3b^4}{12k^5b^2} = \frac{3b^2}{4k^2}$$

$$\frac{30n^2x}{50n^3x^5} = \frac{3}{5nx^4}$$

$$\frac{12t^3a^5}{20t^4a^3} = \frac{3a^2}{5t}$$

$$\frac{3n^2x^5}{9n^4x^6} = \frac{1}{3n^2x}$$

$$\frac{8m^2q^3}{12m^3q^5} = \frac{2}{3mq^2}$$

$$\frac{16w^2y^2}{4w^3y^2} = \frac{4}{w}$$

$$\frac{15t^6w^2}{25t^3w^2} = \frac{3t^3}{5}$$

Simplify each expression.

$\frac{10ky^4}{40k^4y^3} =$

$\frac{30k^4a^6}{20k^4a^3} =$

$\frac{30k^5b^5}{3kb^2} =$

$\frac{15k^3h^3}{3k^2h^2} =$

$\frac{9n^5x^2}{3nx^2} =$

$\frac{8d^2f^4}{4df^3} =$

$\frac{5q^6w^6}{20q^2w^2} =$

$\frac{12y^5n}{3y^3n^5} =$

$\frac{10a^4g^2}{6a^2g} =$

$\frac{20k^5q^5}{10kq^4} =$

$\frac{8n^3b^5}{2n^4b^2} =$

$\frac{3gu^2}{6g^3u^2} =$

$\frac{20t^4k^6}{2t^6k^6} =$

$\frac{8m^6g^2}{2m^5g^3} =$

$\frac{6t^5p}{10t^4p} =$

$\frac{3b^4w^3}{9b^2w^4} =$

$\frac{15b^4q^6}{5b^5q^3} =$

$\frac{20x^3v^3}{10x^5v^2} =$

$\frac{10n^4k^4}{20n^3k^3} =$

$\frac{4w^3q^3}{2w^5q^6} =$

# Answer Key

Simplify each expression.

$$\frac{10ky^4}{40k^4y^3} = \frac{y}{4k^3}$$

$$\frac{30k^5b^5}{3kb^2} = 10k^4b^3$$

$$\frac{9n^5x^2}{3nx^2} = 3n^4$$

$$\frac{5q^6w^6}{20q^2w^2} = \frac{q^4w^4}{4}$$

$$\frac{10a^4g^2}{6a^2g} = \frac{5a^2g}{3}$$

$$\frac{8n^3b^5}{2n^4b^2} = \frac{4b^3}{n}$$

$$\frac{20t^4k^6}{2t^6k^6} = \frac{10}{t^2}$$

$$\frac{6t^5p}{10t^4p} = \frac{3t}{5}$$

$$\frac{15b^4q^6}{5b^5q^3} = \frac{3q^3}{b}$$

$$\frac{10n^4k^4}{20n^3k^3} = \frac{nk}{2}$$

$$\frac{30k^4a^6}{20k^4a^3} = \frac{3a^3}{2}$$

$$\frac{15k^3h^3}{3k^2h^2} = 5kh$$

$$\frac{8d^2f^4}{4df^3} = 2df$$

$$\frac{12y^5n}{3y^3n^5} = \frac{4y^2}{n^4}$$

$$\frac{20k^5q^5}{10kq^4} = 2k^4q$$

$$\frac{3gu^2}{6g^3u^2} = \frac{1}{2g^2}$$

$$\frac{8m^6g^2}{2m^5g^3} = \frac{4m}{g}$$

$$\frac{3b^4w^3}{9b^2w^4} = \frac{b^2}{3w}$$

$$\frac{20x^3v^3}{10x^5v^2} = \frac{2v}{x^2}$$

$$\frac{4w^3q^3}{2w^5q^6} = \frac{2}{w^2q^3}$$

Simplify each expression.

$\frac{16v^4q^4}{20v^6q^3} =$

$\frac{10q^2t^6}{8q^4t^4} =$

$\frac{4g^6d^3}{12g^6d^3} =$

$\frac{2w^3f^5}{20w^3f^3} =$

$\frac{3x^5n^3}{30x^2n^3} =$

$\frac{20n^6p^2}{10n^6p^5} =$

$\frac{50y^4b^3}{40yb^3} =$

$\frac{100b^6p^3}{10b^6p^5} =$

$\frac{20f^5g^5}{8f^4g} =$

$\frac{5r^2b}{10r^5b^3} =$

$\frac{20tk^4}{4t^4k^2} =$

$\frac{5dv^3}{20d^3v^4} =$

$\frac{2f^5m^4}{20f^3m^3} =$

$\frac{25b^5t^3}{5b^4t} =$

$\frac{16u^2n^4}{12u^6n^6} =$

$\frac{20p^6x^3}{2p^2x^2} =$

$\frac{20g^3n^5}{12g^4n} =$

$\frac{40a^5h^2}{10a^2h^2} =$

$\frac{5vf^5}{25v^6f^5} =$

$\frac{20f^4k^3}{15f^3k^5} =$

# Answer Key

Simplify each expression.

$$\frac{16v^4q^4}{20v^6q^3} = \frac{4q}{5v^2}$$

$$\frac{10q^2t^6}{8q^4t^4} = \frac{5t^2}{4q^2}$$

$$\frac{4g^6d^3}{12g^6d^3} = \frac{1}{3}$$

$$\frac{2w^3f^5}{20w^3f^3} = \frac{f^2}{10}$$

$$\frac{3x^5n^3}{30x^2n^3} = \frac{x^3}{10}$$

$$\frac{20n^6p^2}{10n^6p^5} = \frac{2}{p^3}$$

$$\frac{50y^4b^3}{40yb^3} = \frac{5y^3}{4}$$

$$\frac{100b^6p^3}{10b^6p^5} = \frac{10}{p^2}$$

$$\frac{20f^5g^5}{8f^4g} = \frac{5fg^4}{2}$$

$$\frac{5r^2b}{10r^5b^3} = \frac{1}{2r^3b^2}$$

$$\frac{20tk^4}{4t^4k^2} = \frac{5k^2}{t^3}$$

$$\frac{5dv^3}{20d^3v^4} = \frac{1}{4d^2v}$$

$$\frac{2f^5m^4}{20f^3m^3} = \frac{f^2m}{10}$$

$$\frac{25b^5t^3}{5b^4t} = 5bt^2$$

$$\frac{16u^2n^4}{12u^6n^6} = \frac{4}{3u^4n^2}$$

$$\frac{20p^6x^3}{2p^2x^2} = 10p^4x$$

$$\frac{20g^3n^5}{12g^4n} = \frac{5n^4}{3g}$$

$$\frac{40a^5h^2}{10a^2h^2} = 4a^3$$

$$\frac{5vf^5}{25v^6f^5} = \frac{1}{5v^5}$$

$$\frac{20f^4k^3}{15f^3k^5} = \frac{4f}{3k^2}$$

Simplify each expression.

$\frac{20pd^2}{4pd^3} =$

$\frac{6u^2n^4}{10un^5} =$

$\frac{20d^2g^5}{30d^6g^6} =$

$\frac{6bv^5}{9b^2v^2} =$

$\frac{5y^4p^3}{25y^2p^2} =$

$\frac{5n^3u^4}{20n^2u^4} =$

$\frac{8k^5f^4}{2k^2f} =$

$\frac{12w^4n^6}{20w^4n^6} =$

$\frac{50v^2y^5}{5v^3y} =$

$\frac{15x^2t^4}{20x^6t^2} =$

$\frac{8g^5k}{2g^6k^2} =$

$\frac{16uy^6}{4u^2y^2} =$

$\frac{6v^6r^5}{9v^3r^2} =$

$\frac{30x^6t^6}{10x^6t} =$

$\frac{2a^3u^4}{20a^3u^3} =$

$\frac{20d^4q^3}{4d^5q^5} =$

$\frac{4d^4q^6}{40d^6q^3} =$

$\frac{40b^2k}{30b^4k^5} =$

$\frac{2r^3t^3}{6r^4t^5} =$

$\frac{3q^2t^2}{15qt^5} =$

# Answer Key

Simplify each expression.

$$\frac{20pd^2}{4pd^3} = \frac{5}{d}$$

$$\frac{6u^2n^4}{10un^5} = \frac{3u}{5n}$$

$$\frac{20d^2g^5}{30d^6g^6} = \frac{2}{3d^4g}$$

$$\frac{6bv^5}{9b^2v^2} = \frac{2v^3}{3b}$$

$$\frac{5y^4p^3}{25y^2p^2} = \frac{y^2p}{5}$$

$$\frac{5n^3u^4}{20n^2u^4} = \frac{n}{4}$$

$$\frac{8k^5f^4}{2k^2f} = 4k^3f^3$$

$$\frac{12w^4n^6}{20w^4n^6} = \frac{3}{5}$$

$$\frac{50v^2y^5}{5v^3y} = \frac{10y^4}{v}$$

$$\frac{15x^2t^4}{20x^6t^2} = \frac{3t^2}{4x^4}$$

$$\frac{8g^5k}{2g^6k^2} = \frac{4}{gk}$$

$$\frac{16uy^6}{4u^2y^2} = \frac{4y^4}{u}$$

$$\frac{6v^6r^5}{9v^3r^2} = \frac{2v^3r^3}{3}$$

$$\frac{30x^6t^6}{10x^6t} = 3t^5$$

$$\frac{2a^3u^4}{20a^3u^3} = \frac{u}{10}$$

$$\frac{20d^4q^3}{4d^5q^5} = \frac{5}{dq^2}$$

$$\frac{4d^4q^6}{40d^6q^3} = \frac{q^3}{10d^2}$$

$$\frac{40b^2k}{30b^4k^5} = \frac{4}{3b^2k^4}$$

$$\frac{2r^3t^3}{6r^4t^5} = \frac{1}{3rt^2}$$

$$\frac{3q^2t^2}{15qt^5} = \frac{q}{5t^3}$$

Simplify each expression.

$\frac{6w^2v^4}{2w^6v^2} =$

$\frac{2x^2p^5}{6xp^4} =$

$\frac{15h^3b^6}{9h^6b^2} =$

$\frac{30t^4x}{20t^5x^6} =$

$\frac{4hp^3}{6h^5p^2} =$

$\frac{5n^4h^3}{25n^2h^2} =$

$\frac{16y^5b^4}{12y^6b^2} =$

$\frac{20vq}{8v^5q^5} =$

$\frac{5b^4x^2}{20b^4x^3} =$

$\frac{8g^5q^2}{2g^6q^3} =$

$\frac{5v^6t^4}{25vt} =$

$\frac{100m^4n^4}{10m^4n^2} =$

$\frac{50f^3p^6}{20fp^3} =$

$\frac{8u^6r}{2u^3r^5} =$

$\frac{8n^5g^2}{4ng^3} =$

$\frac{6x^5h}{4x^3h^3} =$

$\frac{30t^4w^4}{10t^2w^2} =$

$\frac{15n^2b^4}{9n^3b^3} =$

$\frac{10u^6h}{4u^4h^2} =$

$\frac{4h^3p^5}{10h^4p^6} =$

# Answer Key

Simplify each expression.

$\frac{6w^2v^4}{2w^6v^2} = \frac{3v^2}{w^4}$

$\frac{2x^2p^5}{6xp^4} = \frac{xp}{3}$

$\frac{15h^3b^6}{9h^6b^2} = \frac{5b^4}{3h^3}$

$\frac{30t^4x}{20t^5x^6} = \frac{3}{2tx^5}$

$\frac{4hp^3}{6h^5p^2} = \frac{2p}{3h^4}$

$\frac{5n^4h^3}{25n^2h^2} = \frac{n^2h}{5}$

$\frac{16y^5b^4}{12y^6b^2} = \frac{4b^2}{3y}$

$\frac{20vq}{8v^5q^5} = \frac{5}{2v^4q^4}$

$\frac{5b^4x^2}{20b^4x^3} = \frac{1}{4x}$

$\frac{8g^5q^2}{2g^6q^3} = \frac{4}{gq}$

$\frac{5v^6t^4}{25vt} = \frac{v^5t^3}{5}$

$\frac{100m^4n^4}{10m^4n^2} = 10n^2$

$\frac{50f^3p^6}{20fp^3} = \frac{5f^2p^3}{2}$

$\frac{8u^6r}{2u^3r^5} = \frac{4u^3}{r^4}$

$\frac{8n^5g^2}{4ng^3} = \frac{2n^4}{g}$

$\frac{6x^5h}{4x^3h^3} = \frac{3x^2}{2h^2}$

$\frac{30t^4w^4}{10t^2w^2} = 3t^2w^2$

$\frac{15n^2b^4}{9n^3b^3} = \frac{5b}{3n}$

$\frac{10u^6h}{4u^4h^2} = \frac{5u^2}{2h}$

$\frac{4h^3p^5}{10h^4p^6} = \frac{2}{5hp}$

Simplify each expression.

$\frac{50a^4r^6}{10a^4r^3} =$

$\frac{20r^2w^5}{4r^2w^4} =$

$\frac{100f^3p^3}{10fp^4} =$

$\frac{3u^3h^4}{6u^5h^3} =$

$\frac{20b^6y^5}{50b^5y^2} =$

$\frac{3v^2n^2}{15v^3n^4} =$

$\frac{25v^4a^5}{20v^3a^3} =$

$\frac{40b^6w^4}{10b^2w^3} =$

$\frac{3u^2v^3}{6u^6v} =$

$\frac{40y^5a}{50y^2a^6} =$

$\frac{3h^3y^4}{15h^4y^5} =$

$\frac{3u^2y}{6u^4y^4} =$

$\frac{30t^3w^6}{50t^4w^6} =$

$\frac{16u^5y^3}{20u^6y^4} =$

$\frac{3g^5w}{12g^6w^2} =$

$\frac{20p^2u^4}{15p^2u^5} =$

$\frac{5h^4a^6}{50h^3a^4} =$

$\frac{20p^5b^2}{4pb^4} =$

$\frac{8d^4u^5}{2du} =$

$\frac{10p^2w^3}{25pw^2} =$

# Answer Key

Simplify each expression.

$\frac{50a^4r^6}{10a^4r^3} = 5r^3$

$\frac{20r^2w^5}{4r^2w^4} = 5w$

$\frac{100f^3p^3}{10fp^4} = \frac{10f^2}{p}$

$\frac{3u^3h^4}{6u^5h^3} = \frac{h}{2u^2}$

$\frac{20b^6y^5}{50b^5y^2} = \frac{2by^3}{5}$

$\frac{3v^2n^2}{15v^3n^4} = \frac{1}{5vn^2}$

$\frac{25v^4a^5}{20v^3a^3} = \frac{5va^2}{4}$

$\frac{40b^6w^4}{10b^2w^3} = 4b^4w$

$\frac{3u^2v^3}{6u^6v} = \frac{v^2}{2u^4}$

$\frac{40y^5a}{50y^2a^6} = \frac{4y^3}{5a^5}$

$\frac{3h^3y^4}{15h^4y^5} = \frac{1}{5hy}$

$\frac{3u^2y}{6u^4y^4} = \frac{1}{2u^2y^3}$

$\frac{30t^3w^6}{50t^4w^6} = \frac{3}{5t}$

$\frac{16u^5y^3}{20u^6y^4} = \frac{4}{5uy}$

$\frac{3g^5w}{12g^6w^2} = \frac{1}{4gw}$

$\frac{20p^2u^4}{15p^2u^5} = \frac{4}{3u}$

$\frac{5h^4a^6}{50h^3a^4} = \frac{ha^2}{10}$

$\frac{20p^5b^2}{4pb^4} = \frac{5p^4}{b^2}$

$\frac{8d^4u^5}{2du} = 4d^3u^4$

$\frac{10p^2w^3}{25pw^2} = \frac{2pw}{5}$

Simplify each expression.

$\frac{8r^3x}{2r^2x^6} =$

$\frac{10n^6a^2}{15na^2} =$

$\frac{3f^5y^2}{12f^6y^3} =$

$\frac{9f^6h}{15f^4h^2} =$

$\frac{9y^2d^3}{6y^3d^3} =$

$\frac{12k^4g^3}{8k^2g^4} =$

$\frac{3h^5x^4}{9h^3x^5} =$

$\frac{15dw^5}{10dw^5} =$

$\frac{6d^5v}{3dv^2} =$

$\frac{25q^4m^3}{10qm} =$

$\frac{20wh^2}{12w^5h^2} =$

$\frac{25x^3w^5}{20x^2w^5} =$

$\frac{15m^2v^3}{25m^6v^6} =$

$\frac{12w^6v}{15w^2v} =$

$\frac{15x^4t^2}{25x^5t^4} =$

$\frac{15a^5u}{10a^5u^6} =$

$\frac{40w^3q^6}{4w^6q} =$

$\frac{12d^6t^6}{8d^3t^3} =$

$\frac{6an^6}{2a^2n} =$

$\frac{4k^2r^4}{40kr^4} =$

# Answer Key

Simplify each expression.

$$\frac{8r^3x}{2r^2x^6} = \frac{4r}{x^5}$$

$$\frac{10n^6a^2}{15na^2} = \frac{2n^5}{3}$$

$$\frac{3f^5y^2}{12f^6y^3} = \frac{1}{4fy}$$

$$\frac{9f^6h}{15f^4h^2} = \frac{3f^2}{5h}$$

$$\frac{9y^2d^3}{6y^3d^3} = \frac{3}{2y}$$

$$\frac{12k^4g^3}{8k^2g^4} = \frac{3k^2}{2g}$$

$$\frac{3h^5x^4}{9h^3x^5} = \frac{h^2}{3x}$$

$$\frac{15dw^5}{10dw^5} = \frac{3}{2}$$

$$\frac{6d^5v}{3dv^2} = \frac{2d^4}{v}$$

$$\frac{25q^4m^3}{10qm} = \frac{5q^3m^2}{2}$$

$$\frac{20wh^2}{12w^5h^2} = \frac{5}{3w^4}$$

$$\frac{25x^3w^5}{20x^2w^5} = \frac{5x}{4}$$

$$\frac{15m^2v^3}{25m^6v^6} = \frac{3}{5m^4v^3}$$

$$\frac{12w^6v}{15w^2v} = \frac{4w^4}{5}$$

$$\frac{15x^4t^2}{25x^5t^4} = \frac{3}{5xt^2}$$

$$\frac{15a^5u}{10a^5u^6} = \frac{3}{2u^5}$$

$$\frac{40w^3q^6}{4w^6q} = \frac{10q^5}{w^3}$$

$$\frac{12d^6t^6}{8d^3t^3} = \frac{3d^3t^3}{2}$$

$$\frac{6an^6}{2a^2n} = \frac{3n^5}{a}$$

$$\frac{4k^2r^4}{40kr^4} = \frac{k}{10}$$

Simplify each expression.

$\frac{20w^2n}{50w^2n^4} =$

$\frac{3u^6f^3}{30u^5f} =$

$\frac{8f^4q}{12f^2q^5} =$

$\frac{25a^3v^4}{5a^4v^3} =$

$\frac{4r^5v^3}{6r^6v^3} =$

$\frac{6a^3x^5}{4a^4x^3} =$

$\frac{50y^5a^2}{40y^6a} =$

$\frac{8g^4x^2}{4g^5x} =$

$\frac{6rh^4}{3rh} =$

$\frac{10b^3t^4}{20bt^4} =$

$\frac{5p^3v^4}{50p^5v^5} =$

$\frac{20v^3t^4}{8vt^6} =$

$\frac{10g^3r^5}{20g^5r^5} =$

$\frac{20n^2t}{12n^3t^5} =$

$\frac{50x^2f}{30x^6f} =$

$\frac{25r^3g^3}{15r^5g^5} =$

$\frac{15xw}{10x^6w^3} =$

$\frac{6va^4}{15v^4a^3} =$

$\frac{20r^3q^4}{2r^5q^3} =$

$\frac{4t^4q}{20t^6q^6} =$

# Answer Key

Simplify each expression.

$$\frac{20w^2n}{50w^2n^4} = \frac{2}{5n^3}$$

$$\frac{3u^6f^3}{30u^5f} = \frac{uf^2}{10}$$

$$\frac{8f^4q}{12f^2q^5} = \frac{2f^2}{3q^4}$$

$$\frac{25a^3v^4}{5a^4v^3} = \frac{5v}{a}$$

$$\frac{4r^5v^3}{6r^6v^3} = \frac{2}{3r}$$

$$\frac{6a^3x^5}{4a^4x^3} = \frac{3x^2}{2a}$$

$$\frac{50y^5a^2}{40y^6a} = \frac{5a}{4y}$$

$$\frac{8g^4x^2}{4g^5x} = \frac{2x}{g}$$

$$\frac{6rh^4}{3rh} = 2h^3$$

$$\frac{10b^3t^4}{20bt^4} = \frac{b^2}{2}$$

$$\frac{5p^3v^4}{50p^5v^5} = \frac{1}{10p^2v}$$

$$\frac{20v^3t^4}{8vt^6} = \frac{5v^2}{2t^2}$$

$$\frac{10g^3r^5}{20g^5r^5} = \frac{1}{2g^2}$$

$$\frac{20n^2t}{12n^3t^5} = \frac{5}{3nt^4}$$

$$\frac{50x^2f}{30x^6f} = \frac{5}{3x^4}$$

$$\frac{25r^3g^3}{15r^5g^5} = \frac{5}{3r^2g^2}$$

$$\frac{15xw}{10x^6w^3} = \frac{3}{2x^5w^2}$$

$$\frac{6va^4}{15v^4a^3} = \frac{2a}{5v^3}$$

$$\frac{20r^3q^4}{2r^5q^3} = \frac{10q}{r^2}$$

$$\frac{4t^4q}{20t^6q^6} = \frac{1}{5t^2q^5}$$

Made in the USA
Las Vegas, NV
08 February 2024